SUPER-HIGH-STRENGTH HIGH PERFORMANCE CONCRETE

SUPER-HIGH-STRENGTH HIGH PERFORMANCE CONCRETE

Pu Xincheng

Translated by
Ding Jixin
Neil Milestone

CRC Press
Taylor & Francis Group
Boca Raton London New York

CRC Press is an imprint of the
Taylor & Francis Group, an **informa** business

CRC Press
Taylor & Francis Group
6000 Broken Sound Parkway NW, Suite 300
Boca Raton, FL 33487-2742

First issued in paperback 2017

Version Date: 20120706

ISBN 13: 978-1-138-07564-1 (pbk)
ISBN 13: 978-0-415-51715-7 (hbk)

Library of Congress Cataloging-in-Publication Data

Pu, Xincheng, 1931-
Super-high-strength high performance concrete / Pu Xincheng.
p. cm.
"A CRC title."
Includes bibliographical references and index.
ISBN 978-0-415-51715-7 (alk. paper)
1. High strength concrete. 2. Concrete construction. I. Title.

TA439.P896 2013
620.1'36--dc23 2012015460

Visit the Taylor & Francis Web site at
http://www.taylorandfrancis.com

and the CRC Press Web site at
http://www.crcpress.com

Contents

Foreword

Concrete is the most widely used artificial material in the world today. It will be the main building material in the 21st century due to its exceptional advantages such as the abundant supply of raw materials, low cost, simple manufacture, easy forming, rather high durability, good fire resistance, low cost of maintenance, and so forth. In 2003, the volume of cement output in China reached 825 million tons and that of concrete produced reached 1.5 billion m^3 making these outputs the highest in the world. However, the concrete used in China is a rather low grade. Of the total volume of concrete used, it was for the most part C40 and lower, comprising 89.72%; in recent years, the volume of C45–C60 concrete increased year by year, but still represented only 9.39% of the total volume in 2003; C70–C100 concrete only applied in smaller amounts and as just a beginning; as for C110–C150 super-high-strength concrete, there was no real application. Therefore, the development of concrete science and technology in China is at a rather underdeveloped level.

Super-high-strength high performance concrete has a series of advantages, such as high specific strength, high bearing capacity, low resource and energy consumption, and excellent durability, which can meet the requirements of civil and construction engineering for light weight, high-rise construction, large span, heavy load bearing, and durability that are the trends in concrete science and technology development.

In the past, the study of super-high-strength high performance concrete was seldom pursued in China. In the last 10 years, Professor Pu Xincheng and his collaborators, under the support from the National Natural Science Foundation of China and the Doctoral Foundation of the Ministry of Education of China, carried out a long-term, successful study on super-high-strength high performance concrete. They prepared super-high-strength high performance concrete with a strength of 100–165 MPa and excellent fluidity from ordinary raw materials and common technology. A systematic study of the different aspects of this kind of concrete was carried out, including approaches and basic principles of its manufacture, preparation technology, parameters for mix proportion, present state of application, prospects for its development, and so forth. Some significant successes were achieved.

The addition of efficient active mineral additives is a necessary technical measure for preparing super-high-strength high performance concrete. The analysis method for the specific strength of cement content with the pozzolanic effect of active mineral additive in concrete suggested by Pu Xincheng can be used not only for accurate determination of the activity of various active additives, different sources, and a comparison of the degree of activity, but also for quantitative analysis of the procedure and behavior of the pozzolanic effect, offering a tool for analysis of the composition of concrete strength.

To overcome the high brittleness of super-high-strength high performance concrete, Pu Xincheng and his colleagues combined super-high-strength high performance concrete with steel tubes and made a composite of steel tubes and super-high-strength concrete with high load-bearing capacity and excellent deformation performance, which can be used for constructing a super class building about a kilometer high (referred to by the author as *kilometer compressive material*). It is expected that such steel tube, super-high-strength concrete composite, which gives full play to the advantages of both steel and super-high-strength concrete while overcoming their shortcomings, will promote the development of building structural materials, construction science, and technology.

Super-High-Strength High Performance Concrete summarizes the research achievements in the field of super-high-strength high performance concrete of the author and his collaborators over the last 10 years and is a work of systematic thoroughness. In this book, the author presents his unique understanding of many problems related to such high-strength material.

Tang Mingshu
Academician of the Chinese Academy of Engineering

Preface

In 1992, the National Natural Science Foundation Committee issued a guide for application to the 1993 National Natural Science Foundation, whereby a key project for the eighth 5-year plan, titled "Research on Structure and Mechanical Performances of Super-High-Strength Concrete," was listed and a study on 90- to 100-MPa super-high-strength concrete was required. The Foundation Committee, Ministry of Railway, Ministry of Construction, and the National Bureau of Building Materials jointly sponsored this project with total funding of RMB 1.1 million yuan. The author (his affiliation at that time was with Chongqing Jianzhu University) participated in this project. Subsequently, he took part in the project's defense. Finally, Tsinghua University (the leading institution on the project), the Research Institute of Railway, the China Research Institute of Materials, and Chongqing Jianzhu University won this project jointly and it was divided into several parts: The author was in charge of the part to study the highest strength of concrete (90 to 110 MPa) with funding of RMB 0.2 million yuan. One year later, this project was renamed *Research on Structure and Mechanical Performances of High-Strength and High-Performance Concrete* (59338120). In 1999, the author won a doctoral program-funded project from the Ministry of Education with total funding of RMB 40,000 yuan. With the support of these two projects, the author and his colleagues succeeded after great effort in producing 90–150 MPa super-high-strength and high performance (SHSHP) concrete. Their research work covered different aspects of SHSHP concrete and some successes and important progress were achieved. This book is based on the results of these two projects and review of the related literature data.

In this book, the close relationship between modern concrete and modern civil engineering is described in brief; the main direction for further development of concrete science and technology, that is, pursuing high strength and super-high strength as well as durability and super durability, is set forth; the preparation approach and principles of SHSHP concrete are expanded upon; the composition and strength performance of calcium silicate hydrates are described in detail; the effect of the pozzolanic reaction of active mineral additive on the strength and composition improvement of calcium silicate hydrates and other effects of active mineral additives are presented; the strength composition of SHSHP concrete is analyzed with some examples; the method for preparation of SHSHP concrete with ordinary materials and normal technology is discussed; and the influence of raw materials and mix proportion parameters on the strength and flowability of SHSHP concrete is detailed. In order to help readers understand SHSHP concrete deeply from the structural level, in this book, the composition characteristics of the hydrates of SHSHP concrete, their excellent pore structure, perfect interface structure, and homogeneous macrostructure are described in detail. The strength and deformation performance of SHSHP concrete are the basic properties for their application. In this book, the relationships between the compressive strength of SHSHP concrete and its splitting strength, flexural strength, axial compressive strength, ultimate deformation, and deformation modulus are discussed and the corresponding regression formulas are presented; the character of the stress–strain curve of SHSHP concrete is demonstrated and the high brittleness of such concrete is described. Shrinkage is a basic property of concrete and the self-shrinkage of SHSHP concrete with its low water/cement ratio increases significantly due to internal self-drying caused by the hydration of the cement, leading to an increase in possible shrinkage cracking, which is not an unsolvable problem. This book describes how the shrinkage of SHSHP concrete can be decreased in great degree by diligent, on-time water curing and the addition of an expanding and a shrinkage reducing agent.

The author discusses the excellent durability of SHSHP concrete with its durability indexes; then these indexes are compared with those of ordinary and high strength concretes. The author suggests that the experimental methods based on ordinary concrete are not well adapted to the experiments for SHSHP concrete; they should be improved or new experimental methods should be set up according to the characteristics of SHSHP concrete. Finally, new uses and practical applications for SHSHP concrete are explored. The economic benefits of SHSHP concrete application are verified. Due to the high brittleness of SHSHP concrete, the author presents the optimal method for its application: in a composite material comprised of super-high-strength concrete confined by super-high-strength steel tubes. In practice, a steel tube, super-strength concrete with the strength of 164.9 MPa has been produced, which has excellent deformation performance and quite high-bearing capacity: A kilometer super-high building could be built with such material. Therefore, the author calls such material *kilometer compressive material*. The bright future for the application of SHSHP concrete and kilometer compressive material is discussed.

Little research has been conducted on SHSHP concrete in the world. In China, our research work is, in general, just a start. There are many problems still not studied both at home and abroad, especially SHSHP concrete applications for structural engineering. The author hopes that more scientific and technical professionals become involved in this research field to promote a deeper understanding of SHSHP concrete and its widespread use in engineering.

Involved in these projects were Professor Yan Wu-nan, Associate Professor Li Li-ren, and the author's former students: Professor Wang Zhi-jun (post-doctoral scholar), Lecturers Wang Cong (PhD), Wang Yong-wei (PhD), Wan Chao-jun (PhD), and Professor Tan Ke-feng (PhD). Participants in experimental work were engineer Bai Guang, engineer Pu Huai-jing, and laboratory technician He Gui. The author acknowledges and greatly appreciates the collaborators for their hard work and success in this research. Special acknowledgment is given to Professor Chen Jian-xiong, Research Fellow Lan Hai for offering references in Japanese and German, and engineer Zhao Jun from Beijing of Eken Company, Norway, for the valuable information.

Due to the limitations of the author's academic work, there may be some omissions or errors. Suggestions from specialists both at home and abroad are welcome.

Pu Xincheng

Acknowledgments

The author wishes to thank Professor Ding Jixin of Chongqing University, China for his translation of this book from Chinese into English and Professor Neil Milestone of Milestone and Associates Ltd., New Zealand for polishing the language of the English edition. The author also wishes to thank Taylor & Francis Group for their encouragement and editorial support. Thank you is also extended to Professor Wan Chaojun of Chongqing University, China for his work as an intermediary between Taylor & Francis Group and Chongqing University Press and for his redrawing and beautifying the figures in this book.

Pu Xincheng

About the author

Pu Xincheng is a professor and doctoral advisor at Chongqing University. He is a famous scholar in the field of cement and concrete science in China where he has been engaged for a long time in scientific research and teaching, and his work has attained tremendous success. His research results have been evaluated at the top level throughout the world, and he has made several pioneering and fundamental contributions in a number of fields. In China, he was the first to develop rapid-hardening, high-strength, high-impermeability, and high-durability alkali slag cement and concrete using energy saving and waste utilization methods. Professor Pu Xincheng developed lime-cement-sand-gas concrete and, under his guidance, the first manufacturing of line of lime-sand-gas concrete in China was designed and set up with the capacity of 50,000 m^3 per year. He also made the first successful study of contact hardening cementitious material in China. He was the first in China to develop super-high-strength high performance concrete with excellent flowability and a compressive strength as high as 100–150 MPa using ordinary raw materials and common technology. He also carried out a comprehensive study on the strength, deformation, and durability of this type of concrete. By combining the steel tubing with 165-MPa super-high-strength concrete he was the first in the world to develop kilometer compressive material capable for construction of super skyscrapers more than 1000 m high. He has suggested a method for analyzing the pozzolanic effect of active mineral additives in cement and concrete, which is more advanced and scientific in comparison to existing methods, both at home and abroad.

Professor Pu Xincheng has published two books, three translated works, and more than 190 papers; he has trained 2 postgraduate, 13 PhD, and 21 master's students.

Abstract

This book discusses the technical approach toward the manufacture and the basic principles of 100–150 MPa super-high-strength high performance concrete with common raw materials and common technology. The manufacturing technology and mix proportions of the concrete are described in detail. The characteristics of pore structure, interface structure, and macrostructure as well as the composition character of the hydration products for super-high-strength high performance concrete are studied; the strength, deformation, and durability of such concrete are discussed. The application and technical–economic benefits of super-high-strength high performance concrete currently available are reviewed and the prospects for their application are described.

This book is oriented toward researchers of building materials, professionals and students of institutions of higher education, staff at ready-mix concrete companies, concrete elements manufacturers, and engineers engaged in civil engineering design and construction.

Chapter 1

Modern civil engineering and modern concrete

1.1 CURRENT DEVELOPMENTS IN CIVIL ENGINEERING

From ancient times until now, the advance of human civilization has been closely tied to the development of civil engineering: for the settlement of people, a house should be built; for food supply, irrigation facilities should be built; for clothing, apart from cotton and hemp cultivation, mulberry planting, and silkworm breeding, a textile manufacturing factory should be built; and especially, for the development of modern artificial fibers, large-scale construction of a factory for chemical fiber synthesis is needed; for communication and goods transportation, railways, highways, harbors, and airports should be constructed. Obviously, civil engineering has a close relationship with human life and national economy.

A large number of famous civil engineering projects in China have become part of the world's heritage, representing the culture and progress of various Chinese dynasties such as Dujiang Dam near Chengdu; the Great Wall outside Beijing; the Beijing–Hangzhou Grand Canal starting at Beijing and passing through Tianjin and the provinces of Hebei, Shandong, Jiangsu, and Zhejiang to the city of Hangzhou; Zhaozhou Bridge in Hebei Province; the wooden pagoda of Fogong Temple in Ying County; and the Imperial Palace in Beijing. However, in all of these projects only soil, stone, brick, and timber were used as building materials.

Since the first application of steel and concrete in civil engineering in the mid-1800s and the development of prestressed concrete technology later in the 1920s, the volume of steel and concrete applied in civil engineering has increased rapidly. With the expansion of civil engineering, and an upgrade of the level of science and technology, through continuous development of engineering mechanics, structural science, and the science of building materials, progress in civil engineering has led to a breakthrough clearly illustrated by the construction of skyscrapers, super-large-span bridges, and immense water control projects which require large amounts of work and high levels of technology.

Table 1.1 Height of the highest buildings at different periods of time

Year	*Name of the building*	*Number of stories*	*Height (m)*
1870	New York Equitable Life Insurance Building	7	25
1885	Chicago Home Life Insurance Company Building	10	55
1913	New York Woolworth Building	59	241
1930	New York Chrysler Building	77	319
1931	New York Empire State Building	102	381
1973	New York World Trade Center	110	417
1974	Chicago Sears Tower	109	442
1997	Petronas Towers, Kuala Lumpur, Malaysia	88	450
2003	Taipei International Financial Center, China	101	508
2009	Burj Khalifa Tower (Burj Dubai), Dubai, United Arab Emirates[a]	168	828

[a] Supplemented in 2011.

1.1.1 High-rise buildings and skyscrapers

With enlargement of the scale of city development and population growth, the rise in land cost, the progress of construction technology, and the rapid growth of the economy, buildings are becoming larger and taller. In Table 1.1, the height of the highest buildings represented in different periods is given. In 1870, the Equitable Life Insurance Building built in New York City had only seven stories with a height of about 25 m. In 1997, the Petronas Building built in Kuala Lumpur was as high as 450 m; then in 2003, the Taipei International Financial Center in China reached 508 m, with the height of the built structure at 455 m exceeding the Petronas Towers, making it the highest building in the world in 2008. Most recently, Burj Dubai in Dubai, United Arab Emirates, was topped out at 828 m (2717 ft) on January 17, 2009, but it is yet to be completed. In China during the 1950s, the highest building was the Shanghai Park Hotel (24 stories) and now many high-rise buildings of more than 30 stories can be found in all large cities of the country. In 1998, the Jinmao Building in Shanghai, 88 stories and 420 m high, was built and it was the highest on the Chinese Mainland, but the Shanghai World Financial Center, 101 stories and 492 m high, is taller.

1.1.2 Large span and super-large-span bridges

In bridge engineering, the suspension bridge has the ability to span the greatest distance. The longest suspension bridge in the world is currently the Akashi Kaikyo Bridge in Japan (1991 m). In China, suspension bridges with spans of more than 1000 m include the Tsing Ma Bridge in Hong Kong

(1377 m), the Jiangying Yangtze Bridge (1385 m), and the Runyang Yangtze Bridge (1490 m).

Cable-stayed bridges offer only slightly less span with, at present, the longest span belonging to the Tatara Ohashi Bridge in Japan (890 m), followed by the Normandy Bridge in France (856 m), with the Yangpu Bridge in Shanghai (602 m) in third place. The longest arched bridge is the Wanzhou Yangtze Bridge in China built in 1997 with a span of 420 m, followed by that of the KAK-II Bridge in Croatia (390 m) built in 1980. At present, the Wushan Yangtze Bridge is under construction which, with a span of 460 m, will be more than that of the Wanzhou Yangtze Bridge. The recently completed bridge across the Hangzhou Bay connecting Shanghai and Ningbo is the longest bridge in the world (36 km), the construction of which started in 2003 and was completed in 2008.

1.1.3 Dam construction

Dams are constructed for power generation, flood control, water storage, and shipping, and they provide a challenge from the viewpoint of the volume of concrete, investment, technical difficulty, and time to build. During the recent development in China, 86,000 reservoirs from large to small have been completed giving a total volume of water storage of 458 billion m^3 and a hydropower generating capacity of 63 million KW, ranking China third in the world in this domain. Among these are several famous large irrigation and hydropower projects, such as Sanmenxia, Liujiaxia, Xiaolangdi, Ertan, and Gezhouba. The Yangtze Three Gorges Project with water storage up to 139 m^3 was started in 2003 and completed in 2009 with a hydropower capacity of 18.2 million KW, which exceeds the previous highest, the Itaipu Hydroelectric Power Station (12.6 million KW) in Brazil–Paraguay. The volume of concrete poured for the Yangtze Three Gorges Dam reached 28 million m^3, in comparison with that of 13 million m^3 for the Itaipu.

Within China, construction for transportation and urban development has reached new levels with over 68,000 kilometers of railway track laid, the longest in Asia and the third longest in the world. Roading has also increased dramatically. Prior to 1988, there were no expressways on the Chinese Mainland, but by 2003 more than 40,000 kilometers had been constructed.

The importance of civil engineering is self-evident due to its close relationship to the well-being of a country and the daily life of its people. In today's modern world, construction of factories, mines, railways, bridges, irrigation and hydropower projects, shopping centers, schools, and hospitals along with telecommunication facilities are essential infrastructures for society. Securing the capital funds for these are at the forefront of development.

1.2 THE HISTORY OF MODERN CONCRETE DEVELOPMENT AND ITS SIGNIFICANCE IN MODERN CIVIL ENGINEERING

In any civil engineering project, materials form the basis for construction. Historically, natural materials such as soil, stone, and wood were used as building materials with bricks from fired clays playing an important role in medieval Europe. Later, lime and pozzolana were used as a cementing material and mixed with water, sand, and stone to make an artificial stone, which served as a building material. In ancient Rome, many famous constructions such as the Pantheon, the Roman amphitheatre, harbors, lighthouse, and aqueducts were built with such ancient concrete and are still preserved today, 2000 years after their construction. Such concrete has its shortcomings such as slow hardening and low strength, and it was difficult to control the concrete quality due to limitations of the heterogeneous pozzolana resource. Nevertheless, this ancient concrete played an important role in construction in Europe up to 19th century when a new binder appeared.

After the invention of Portland cement in 1824 by Aspdin (it is called *silicate cement* in China), there was a rapid development in its use as the cementing material for concrete. There are abundant resources for manufacture of such cement, which can be made with controllable quality and which hardens rapidly giving high strength. Therefore, concrete technology developed quickly, and the volume used increased as the scope for its application increased. At present, it is the most abundant artificial material in the world.

Concrete has a series of advantages such as an abundant supply of raw materials, low cost, simple manufacturing technology, convenience in forming, durability, low maintenance expenses, and good fire resistance. However, it has one significant weak point, its low tensile strength. Only in the midst of the 19th century, with the development of the steel industry, was steel reinforcement introduced into concrete, forming a new composite material—reinforced concrete, which remedied the defect of pure concrete—insufficient tensile strength. This was a great step forward in the application of concrete in various structures.

The introduction of reinforcement into concrete allowed its use as tensile or bending elements, but the liability for cracking still remains a problem. Applying prestressing to concrete by tensioning the steel reinforcement can ensure the concrete member resists tension under loading so no crack is formed. Especially for high strength materials, the prestressing method is the most effective way of preventing cracking. Development of the theory of shrinkage and creep for concrete, together with the use of high-strength steel wire and anchors, laid the foundation for application of prestressing technology in concrete engineering.

The appearance of prestressed concrete was another step in the development of concrete technology, which improved concrete's properties through changes outside the concrete itself. Owing to excellent results in prestressing technology in large span construction, in high-rise buildings as well as in seismic resistance, where cracking resistance and resistance to inner pressure has been increased, the scope of the application of concrete has expanded greatly, with C100 prestressed concrete construction competing closely with steel on a weight basis. Therefore, prestressed concrete construction can now substitute for most steel construction. Today, the applications of concrete extend from industrial to civil construction, transportation to dams, floating marine and undersea construction, and pressure tanks and containers for natural gas storage and nuclear power stations.

The use and application of expanding cements or addition of expansive agents into concrete to produce shrinkage-compensating concrete or self-stressing concrete is another outstanding success in the development of concrete technology. The essence of this technology is to change the shrinkage property of concrete into one of increasing expansion; to prevent cracking due to its shrinkage, or to tension the steel reinforcement by using its expanding property. Such a modification is based on the combination of external and internal conditions. Expanding concrete is widely used in industrial or civil constructions, road pavement, storage tanks, self-stressed pipe, impermeable and waterproof constructions, and connections for pipes, concrete element joints, secondary grouting, and so forth.

Provided all other conditions remain the same, the strength of concrete primarily depends on the water/cement ratio, the famous water/cement ratio theory. This indicates that under full compaction conditions, the lower the water/cement ratio, that is, the less water added, the higher the strength of the concrete. However, if the water/cement ratio is too low, the fluidity of the mixture decreases rapidly, and it will cause difficulties in compaction, making the concrete less compact, thus leading to a decrease in the concrete strength.

The invention and application of water-reducing agents was an important step in the development of concrete technology over the last 20 to 30 years. Adding a water reducer or superplasticizer into the concrete mixture allowed the water/cement ratio to be reduced without loss of fluidity of the mix providing an increase in strength. Technological processes such as mixing, transportation, placement, and forming of the mix become easier and the properties of the concrete are improved. At present, water-reducing agents are the most widely used admixture in modern concrete engineering due to both their technical and economical advantages.[105]

Although concrete is now a traditional building material, its advantages and the progress in understanding the science and technology of concrete as a material have meant it has become the most widely used artificial material in the world and it is a key material in civil engineering. For example, the

Petronas Towers in Kuala Lumpur, constructed at the end of the 20th century, ranked as the world's second highest building; the Shanghai Jinmao Building, the fourth tallest in the world, and Taipei International Financial Center, the highest in the world,* all have concrete as their main structural material. The annual output of cement in China reached 825 million tons in 2003, and the volume of concrete produced more than 1.5 billion m^3 (where the ready-mixed concrete reached 215.3 million m^3).[1] It follows that concrete has significant importance in civil engineering and national construction.

1.3 HIGH STRENGTH AND SUPER-HIGH STRENGTH OF CONCRETE

Apart from the incomparable advantages mentioned above, concrete has some crucial shortcomings. One of these is its high brittleness, especially since this brittleness increases with strength, demonstrating the low tensile strength to compressive strength ratio. In ordinary concrete, this ratio is about 1:10, the reason being that in the microstructure of concrete, covalent bonds play a leading role, which is difficult to change with raw materials or mix proportions. This crucial property of concrete can be overcome by means of reinforcement with either metal or organic materials making concrete a composite material. Thus, reinforced and prestressed concretes became important. In order to further decrease the brittleness of the concrete and enhance its ductility, homogeneous distribution of reinforcement has been considered and new composites such as fiber-reinforced concrete, steel mesh concrete, 3D fiber-reinforced concrete, and steel tube restricted concrete have emerged. The ductility of such composite materials is increased greatly and is discussed in Section 11.3.

The other shortcoming of the concrete widely used at present is its heavy dead weight and low strength. For example, the apparent density of ordinary concrete is about 2400 kg/m^3, but its strength is about 10 to 40 MPa.[2] This means the specific strength, that is, the strength of a unit weight of the material, is quite low (Table 1.2). As a result, the cross section of a structural element is large and the weight of the structure itself is high, the transportation volume increases and consumption of material and energy increases, leading to an increase in the cost of construction. That is the disadvantage called *huge beam, large column, and deep foundation.* Especially for the high-rise and super-high-rise building or large spanned and super-large-spanned construction in modern civil engineering, such material cannot be adapted to current requirements.

* After April 8, 2008, Burj Dubai (Burj Khalifa Tower) became the highest building in the world.

Table 1.2 Specific strength by weight for steel, wood, ordinary concrete, and super-high-strength concrete

Material	*Apparent density (kg/m³)*	*Compressive strength (MPa)*	*Specific strength by weight*
Low carbon steel[a]	7860	420	0.053
Wood (pine)[a]	500	35 (parallel to grain)	0.070
Ordinary concrete[a]	2400	30	0.012
Super-high-strength concrete	2600	150	0.058
Lightweight aggregate concrete	1400–1900	15–50	0.011–0.026

[a] Zhou Shiqiong, *Building Materials*, Beijing: China Railway Press, 1999 (in Chinese).

There are two ways out. One is to lighten the dead weight of the concrete used, for example, by using lightweight aggregate concrete with an apparent density in the range of 1400 to 1900 kg/m³, but with a compressive strength of 15 to 50 MPa and specific strength of 0.011 to 0.025, though such improvement is not really effective. A lightweight aggregate with excellent performance (e.g., keramsite) is usually made by calcination, making it costly, and production resources are limited. Therefore, the application of lightweight aggregate concrete is limited, both in China and abroad.

The other way is to use high strength or super-high-strength concrete, where the apparent density of the concrete remains constant or slightly increased though the strength is greatly increased. This is a promising approach and has led the author to study super-high-strength concrete and to compile this book. In fact, the development of the science and technology of concrete proceeded just this way.

In the 1950s, the strength grade of concrete used in China was generally C20. This increased in the 1960s to C30 but during this period, a lot of water was added to concrete to ensure good workability so the water/cement ratio reached 0.5 to 0.7. In the 1970s and 1980s, the introduction of a water reducer allowed a reduction in the water/cement ratio to 0.45 to 0.6 and concrete with a strength grade of C40 to C50 was used widely. In the 1980s and 1990s, to obtain even higher strength and at the same time retain good workability of the concrete mix, superplasticizers with higher water-reducing ability were used and the water/cement ratio was lowered gradually to 0.45 to 0.30 or less, and grade C100, super-high-strength concrete was obtained. In practice, the most widely used concrete is C50 to C70 but C80, C90, and C100 concrete are now being used.

The author of this book has prepared super-high-strength concrete with good workability and compressive strength up to 150 MPa with apparent density of no more than 2600 kg/m³, only 200 kg/m³ more than for ordinary concrete, but the strength is five times as high as the 30 MPa of ordinary concrete. Its specific strength is enhanced reaching 0.058 and exceeding that of steel given in Table 1.2. This illustrates the significance of the super-high strength of the concrete.

1.4 DURABILITY AND SUPER DURABILITY OF CONCRETE

The basic shortcoming of ordinary concrete is its record of poor durability. The service life of common concrete is reckoned as 50 to 100 years, but there are many cases where repair or reconstruction was needed after only 10 to 20 years of service. Even for a huge project such as the Channel Tunnel, the designed service life is only 120 years. However, wooden construction often thought of as temporary shows excellent durability due to mature preserving technology, for example, in China where there are many historical temples or palaces built several hundred years ago that still display their brilliance. Therefore, the widely spread concept that concrete itself is a durable material should be corrected.

Today, the total volume of concrete used in a single project in China continues to increase as the country embarks on a massive building program; for example, the expressways for the *five north–south and seven east–west* highway networks, the planned high speed railway network, the Three Gorges project, the North–South Water Transfer Project, and so forth. Such huge projects will become a heavy burden on society if the concrete is of insufficient durability. You can imagine that if the Three Gorges Project, which took several decades for planning and approval with 17 years of construction, is damaged after 50 to 100 years of service and needs reconstruction, this will be a major drain on the future economy. We should be concerned and learn from experiences in America, where according to sources in the United States, the concrete infrastructure is worth about $6000 billion but requires about $300 billion for repair or reconstruction annually, due to insufficient durability of the concrete. Among the 500,000 highway bridges, more than 200,000 have been damaged and needed repair with, on average, 150 to 200 bridges per year collapsing partially or completely after a service life of less than 20 years with replacement cost or repairs costing $90 billion. Now that a large scale of infrastructure construction is being carried out in China, this type of disaster should be avoided.

The super durability of concrete for a project should be determined by its significance, the state of concrete materials, and economic feasibility; except for temporary or small projects, the service life of middle- or large-sized permanent projects should be enhanced to 500 years; for some key projects, more than 1000 years; and for extra-large key projects, more than 2000 years. If this can be fulfilled, large sums for repair or reconstruction expenses over a long period will be released.

Super durability is our expectation but not utopia; such a target can be reached with effort.

The simplest way to enhance the durability of modern Portland cement-based concrete is to improve its microstructure and the composition of cementing materials, detailed by Pu Xincheng.[3] Some researchers have suggested, "For the concrete, durability is of first importance and high strength does not mean high durability" and they even state that "high strength surely does not lead to durability," when they contrast the strength of concrete to its durability. The author suggests that there are times when the durability of high strength concrete is lower than that of low strength concrete. A typical example is air-entrained concrete, where the strength is lower than that of concrete without air entrainment, but the frost resistance of the former is much better. However, in general, the durability of concrete is closely related to its strength. The basis for this relationship relies on the water/cement ratio. It is widely known that for compactable concrete, the lower the water/cement ratio, the lower the porosity of the concrete and the higher the compactness and strength of the concrete, and there is increased impermeability. Figure 1.1 illustrates the relationship between the water/cement ratio and the penetration of a chloride ion.[4]

Figure 1.1 indicates that the lower the water/cement ratio, the less there is penetration of the chloride ion with no penetration of the chloride ion in

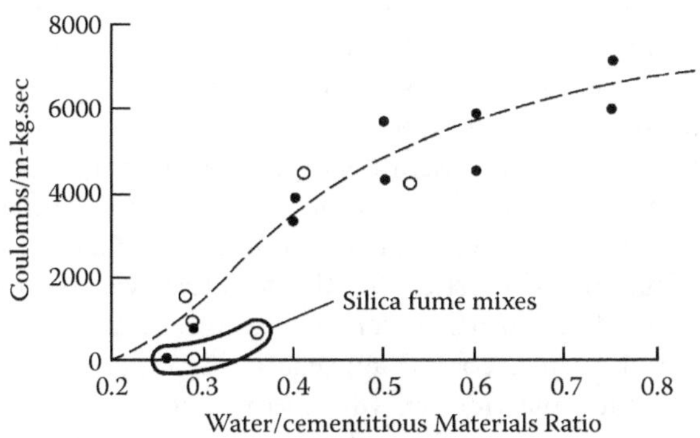

Figure 1.1 Rapid chloride permeability versus water/cementitious materials ratio (water/cement ratio) of concrete. (From Edward G. Nawy, *Fundamentals of High Strength High Performance Concrete*, London: Longman Group Limited, 1996.)

silica fume concrete with a water/cement ratio less than 0.3. Many of the durability indexes of concrete are directly related to the impermeability of the concrete. Therefore, high strength concrete usually has high durability. For super-high-strength concrete, the water/cement ratio is low and compactness is high, so even water and air cannot penetrate it, and therefore the durability indexes are increased greatly. This will be discussed in detail in Chapter 9, "Durability of super-high-strength high performance concrete." Measures do need to be taken to avoid cracking of the concrete and this is possible and discussed later. Consequently, super-high strength and super-high durability of concrete can be obtained simultaneously by taking special precautions.

High durability and super-high durability of concrete can be realized not only through high strength or super-high-strength concrete but also in low strength as well. For example, lime pozzolan cement concrete from ancient Rome is slow setting and has low strength. However, many ancient Roman constructions built from such concrete, such as the Pantheon, Roman amphitheatres, harbors, aqueducts, and baths, are surprisingly still intact and in operation, more than 2000 years after they were built despite the action of natural factors such as running water, rain, snow, and sea water.

The author suggests that to enhance the durability of modern, low strength ordinary concrete based on Portland cement, there are two ways: improvement of its structure and optimization of the hydrate composition. Low cement content should be used with the addition of a mineral additive; the water/cement ratio should be low or medium to reduce porosity and enhance the compactness to avoid the formation of structural defects. The addition of a durability-improving agent[5] is another way to enhance the durability of ordinary concrete. For example, in the new construction of a church in Kobe, Japan, this durability-improving agent has been applied and the expected service life for this construction is now 500 years. Without the measures mentioned above, it is possible to increase the durability of low strength, ordinary Portland cement concrete significantly. For example, for a church built in 1999, on Kauai Island in the United States, the design strength required of the concrete raft foundation was C20 with 1000 years of service life. The mix proportion for the concrete was as follows: cement 106 kg/m^3, fly ash 142 kg/m^3, water 100 kg/m^3, water/cement ratio 0.4, fine aggregate 944 kg/m^3, coarse aggregate 1121 kg/m^3, water-reducing agent 770 ml/m^3, superplasticizer 3480 ml/m^3, and an air-entraining agent 116 ml/m^3.[6] This is a good example of a mix proportion for a low strength concrete where low cement content, a large addition of a mineral additive, and medium to low water/cement ratio provide good durability. Of course, the success of research in this field is not great, and study and more research of this problem are needed.

1.5 MODERN CONCRETE AND THE ENVIRONMENT

With increasing concern for the concept of environmental protection, the production, consumption, or application of any commodity, or construction project should be considered in light of any potential damage to the natural environment. Does it create a hazard to the inhabitants' health? Does it conform to the principle of sustainable development?

As for concrete, the most used building material in the world, does it conform to the principle of sustainable development? Is it an environmentally friendly material? All this attracts the attention of people.

Research shows[7] that during the whole life cycle of concrete, that is, the process from raw material extraction, cement manufacture, concrete preparation, concrete application to recycling, and waste treatment to the recycling for reuse, the following presentation can be given:

1. As observed in Germany for 30 years, during the manufacture of cement, there has been only a trace of toxic pollutant in the flue gas discharged into the atmosphere. Many heavy metal elements are consolidated in the clinker with little discharged with the flue gas. A minor part becomes entrapped, entrained into the soil or water, and there is no adverse effect on the soil and water quality. During application of manufactured concrete, such heavy metals are still consolidated in the concrete without forming a solution. Even at the end of the service life of cement and concrete in concrete buildings, there is little heavy metal leached out in disposal, giving levels much less than those specified in the European standard for drinking water. Therefore, during the whole of the service life of cement and concrete and after their disposal, the amount discharged or leached is small. So, from this aspect, concrete should be considered an environmentally friendly material.
2. Low radioactivity of concrete. Usually, the content of radioactive elements in concrete is extremely low and much lower than that in natural granite and basalt. But attention should be paid when fly ash is used as an additive because fly ash from the burning of some coals leads to radioactivity levels higher than the norm, so the application of such fly ash should be avoided. Thus, with respect to radioactivity levels, concrete is an environmentally friendly material.
3. No radon is emitted from concrete where the concentration of radon in concrete is several hundred times lower than that in soil. In contrast, concrete can absorb and store some radon discharged from the soil. So in terms of radon emission, concrete is an environmentally friendly material.
4. During the manufacture of cement, not only can a null waste discharge be realized but the different kinds of industrial slag or wastes

can be incorporated. For example, the wastes, which can serve as raw material for cement, could be red mud, casting sand, fly ash, sulfuric acid residue, desulfated gypsum, sludge from wastewater treatment plants, and so forth.

Besides, some combustible wastes such as used tires, waste plastic, waste engine oil, waste textiles, waste coating, paint, and solvents can serve as secondary fuel for a cement kiln. Due to the high temperature of the cement kiln (up to 1600°C) and its large thermal capacity, and long residence time for gas and charge inside the kiln, in addition to the basic atmosphere, all hazardous matter in the waste is decomposed or burnt. The slag, ash, and heavy metals are contained within the clinker, with the amount of dioxin and furan discharged from the kiln flue at much less than the standard specification for burnt garbage.

To prepare high strength and super-high-strength concrete, an active mineral additive must be added. Industrial wastes, which can function as active mineral additive to produce high strength and super-high-strength concrete, are fly ash, burnt gange, and silica fume from a silicon smelter. Consequently, a considerable amount of industrial waste can be consumed in the manufacture of cement and concrete and contribute to environmental protection.

However, during the manufacture of Portland cement, the decomposition of $CaCO_3$ and the combustion of fuel discharges a great deal of gas, mainly CO_2, SO_2, and NO_x, into the atmosphere, enhancing the greenhouse effect. In manufacturing 1 ton of clinker, about 1 ton of CO_2 is released to the atmosphere contributing to the rise of atmospheric temperature, causing abnormal weather, melting of the polar ice cover, and a rise in sea level, causing global environmental disaster; this is the unfriendly side of the cement industry.

In order to decrease this unfriendly effect of the cement and concrete industry on the environment, one way would be to use less cement in concrete with greater additions of mineral additives and make full use of the strength potential of the cement by enhancing the contribution of a unit cement content to the strength. Then, the cross-sectional area of the load-bearing unit could be decreased with a concomitant decrease in cement content. This requires production of super-high-strength high performance concrete. For example, when the strength of concrete increases from 30 MPa to 150 MPa, the cement content increases from 300 kg/m^3 to 600 kg/m^3 and the strength contributed per unit cement volume increases from 0.1 MPa/kg to 0.25 MPa/kg, that is, by 150%. In addition, this enhanced strength increase allows the sand, stone, and other materials to play their full role in strength and all materials can be used effectively. The rise in strength and decrease in cross section (see Section 11.2) allow the volume of materials to be reduced significantly and the energy consumption for processing and transportation of these materials to decrease accordingly.

Therefore, development of super-high-strength high performance concrete can reduce the effect the cement industry has on the environment.

One of the basic ways to reduce the CO_2 concentration in the atmosphere is to use plants to absorb CO_2. Apart from raising the amount of forest coverage, another way is to transform a building into a carrier for growing plants, increasing the green area on the earth and increasing the amount of CO_2 absorbed by vegetation. With the high strength of super-high-strength concrete and its high load-bearing capacity, the opportunity exists to transform a high-rise building into an artificial green peak or forest (see Section 11.4), absorbing CO_2 and transforming the concrete indirectly into an environmentally friendly material.

Chapter 2

Approaches and principles

2.1 DEFINITION OF HIGH PERFORMANCE CONCRETE

For more than 20 years, the term *high performance concrete* has been used throughout the world and the number of papers using this term has increased rapidly. The term was needed in concrete engineering and it emphasizes that with the use of high strength concrete in construction comes considerations of high durability and high performance. However, the definition and interpretation of the term *high performance concrete* vary in different engineering fields, in different countries, and for different researchers, so that the requirements are not clear. For example, the American National Institute of Science and Technology (NIST) and American Concrete Institute (ACI) at the 2nd International Symposium on the Utilization of High Strength Concrete, in Berkeley, CA, in the United States, in May 1990 suggested the following:

> High performance concrete is one that is homogeneous with certain properties, for which a strict construction technology should be applied. High quality raw materials should be used, as well as ease of casting without segregation; its mechanical properties should be stable and with high early strength. It is a durable concrete with high stiffness and volume stability. Such concrete is suitable for the high-rise building, bridge and construction working in severe environments.

In 1990, Mehta[8] suggested that

> High performance concrete should have not only high strength, but high durability (resistance to chemical corrosion) and other properties, such as high volume stability (high modulus of elasticity, low dry shrinkage, low creep, and small thermal deformation), high impermeability, and good workability.

Malier[9] from France in a meeting in 1992 defined high performance concrete as

> High performance concrete is characterized by good workability, high strength and sufficient early strength, economical for engineering and high durability. It is especially suitable for key concrete structures such as bridges, harbors, nuclear reactors and expressways.

At the same meeting, Ozawa Ichi and Okamura[10] from Japan suggested that

> High performance concrete should have high workability (high fluidity, cohesion and easy for casting), low temperature rise, low dry shrinkage, high impermeability and sufficient strength.

In a completely different field, Canadian researchers have also called the concrete used for filling disused mines, which has strength of 1 MPa, *high performance concrete.*

Thus, in usage and in the literature, many scholars give their own definition to the term *high performance concrete.*[10,11]

In many of these definitions, an emphasis is placed on aspects of strength, fluidity, and durability, often combining all three. Some researchers consider concrete, which meets some of these requirements, *high performance concrete.* It is easy to understand why this concrete can be used for many projects with different requirements.

All these definitions do reflect the requirements and limits of the quality of the concrete, many of them difficult to quantify. The author feels that of all scientific terms, the definition of high performance concrete is the least strict, and most confusing and difficult to understand. He believes that high performance concrete should meet the following basic requirements:

1. In its fresh state, the concrete should be flowable with good adhesion, it can be placed without segregation or stratification, and it has excellent performance in construction work.
2. The hardened concrete should have high strength. Since strength is the basic property of concrete, ordinary low strength concrete cannot be considered high performance concrete. This does not mean that for ordinary concrete, a good working performance and sufficient durability may be neglected.
3. The hardened concrete should have high volume stability, that is, low shrinkage and low creep.
4. High durability.

It is not reasonable to call concrete *high performance* when it only meets one of these engineering requirements; otherwise, the term *high performance concrete* has too wide a range and the term loses its significance.

In the requirements given above, statements 2, 3, and 4 are not contradictory but are interrelated. In general, the higher the strength, the less the shrinkage and creep, giving rise to a high index of durability. Statement 1 appears to contradict statement 2. To obtain high strength, the water/cement ratio should be low and the water dosage small, whereas high fluidity usually requires a large water dosage and high water/cement ratio. In the mix design of ordinary concrete, the main task is to balance these two requirements in order to meet the requirements for strength and fluidity. Therefore, to prepare high performance concrete with the properties given above, other advanced technological measures are needed, that is, the addition of superplasticizer and effective active mineral additives.

2.2 DETERMINATION OF HIGH PERFORMANCE CONCRETE

As mentioned above, there are four requirements generally accepted for high performance concrete. But which indexes should be used to describe and determine the properties of high performance concrete? On the one hand, there is no single complex index that is capable of describing high performance concrete, as the properties of concrete are multi-aspect. On the other hand, it is difficult to determine high performance concrete using several indexes, which would cause complexity and difficulty. At present, some researchers use a strength index to describe high performance concrete because many properties of concrete are directly related to strength. Pliskin[12] provides an indication based on strength as shown in Table 2.1.

The deficiency of this classification is that it is only made on the index of strength without that of fluidity so that a concrete prepared from a stiff mix with a low water/cement ratio and force compacted could be considered high performance or super-high-performance concrete. Obviously, such concrete cannot meet the first requirement for the high performance concrete given above. Therefore, the author suggests that high performance concrete should be described by at least two indexes, strength and

Table 2.1 Classification of concretes

Type of concrete	*28 d compressive strength (MPa)*
Ordinary concrete (OC)	20–50
High performance concrete (HPC)	50–150
Very-high-performance concrete (VHPC)	100–150
Exceptional concrete (EC)	>150

Source: L. Pliskin, High Performance Concretes—Engineering Properties and Code Aspect, in *High Performance Concrete—From Material to Structure*, London: *E & FN SPON*, 1992: 186.

fluidity, simultaneously. Such an approach is scientific and not complex. However, at present, there is no consensus regarding which index of fluidity high performance concrete should meet. As pumped concrete is now widely used in modern construction work with its high efficiency, low labor costs, and high quality, the author suggests that high performance concrete should have a slump of more than 120 mm, providing concrete that is easy to pump, and that super-high-performance concrete should have a slump of 180 mm, and flows. Using these criteria, the author has classified all concrete into 20 classes in terms of fluidity and strength. As shown in Table 2.2, only the concretes in classes 12 to 20 can really be called *high performance*. The principle behind the different classes is: for fluidity—increasing concrete fluidity uses adjectives such as low flowability, flowable, highly flowable, flowing, highly flowing (or super flowing); for strength—increasing strength is defined as ordinary, high strength, super-high strength, and ultra-super-high strength. The term *common* is identical with our classification, and represents ordinary concrete, along with highly flowable concrete and flowing concrete, and so forth. It should be noted that the level of development of concrete science and technology has not yet reached the point where the values have high accuracy; the methods for determination are not precise although they are reasonable. Therefore, the classes defined by fluidity and strength in Table 2.2 only seem apparent and while defining the category, cannot be taken as absolute.

In Table 2.2, stiff concrete with zero slump is not included (there is a concrete with a stiffness of 700 s), but because such concrete requires powerful equipment for its compaction, it is rarely used.

According to Gjorv,[13] a concrete with a compressive strength as high as 230 MPa can be prepared with high quality, natural mineral aggregate. The author has prepared a concrete with a slump of 220 to 258 mm and compressive strength of 150 to 165 MPa from natural limestone aggregate,[14] which can be used for engineering in the future. Therefore, in Table 2.2, the author has termed concrete with a strength ≥150 MPa *ultra-high-strength concrete.*

2.3 TECHNICAL APPROACHES AND PRINCIPLES FOR PREPARATION OF SUPER-HIGH-STRENGTH HIGH PERFORMANCE CONCRETE

It is well known that for most solid materials, their theoretical strength is (0.1–0.2) × E (where E is the modulus of elasticity of the material) and the real measured strength is (0.1–0.2) × 10^{-3} E. Such a great difference is due to the large amount of structural defects in real materials. Therefore, to prepare high strength materials, first of all, these defects should be eliminated as much as possible.

Table 2.2 Classification and concrete terms suggested by the author

Class	*Term*		*Flowability (slump) (mm)*	*Compressive strength (MPa)*	*Description*
1	Ordinary concrete	Low flowability	1–59	10–49	Cannot be pumped, compacted by vibration, poor durability
2		Flowable	60–119	10–49	Poor pumpability, compacted by vibration, poor durability
3	Highly flowable concrete		120–179	10–49	Suitable for pumping, slight vibration required, poor durability
4	Flowing concrete		180–239	10–49	Suitable for pumping, slight vibration required, poor durability
5	High flowing concrete		≥240	10–49	Suitable for pumping, self-leveling, poor durability
6	Low flowability, high strength concrete		1–59	50–99	Cannot be pumped, compacted by vibration, good durability
7	Flowable, high strength concrete		60–119	50–99	Poor pumpability, compacted by vibration, good durability
8	Low flowability, super-high-strength concrete		1–59	100–149	Cannot be pumped, compacted by vibration, excellent durability
9	Flowable, super-high-strength concrete		60–119	100–149	Cannot or can hardly be pumped, compacted by vibration, excellent durability
10	Low flowability, ultra-high-strength concrete		1–59	≥150	Cannot be pumped, compacted by vibration, excellent durability
11	Flowable, ultra-high-strength concrete		60–119	≥150	Cannot be pumped, compacted by vibration, excellent durability

continued

Table 2.2 (continued) Classification and concrete terms suggested by the author

Class	*Term*		*Flowability (slump) (mm)*	*Compressive strength (MPa)*	*Description*
12	High performance concrete (HPC)	High flowable, high strength concrete	120–179	50–99	Suitable for pumping, compacted by vibration, good durability
13		High flowable, super-high-strength concrete	120–179	100–149	Hardly be pumped, compacted by vibration, excellent durability
14		High flowable, ultra-high-strength concrete	120–179	≥150	Hardly be pumped, compacted by vibration, excellent durability
15		Flowing high strength concrete	180–239	50–99	Suitable for pumping, slight vibration required, good durability
16		High flowing, high strength concrete	≥240	50–99	Suitable for pumping, vibration required, good durability
17	Super-high-strength HPC	Flowing super-high-strength concrete	180–239	100–149	Can be pumped, vibration required, excellent durability
18		High flowing, super-high-strength concrete	≥240	100–149	Can be pumped, self-leveling, excellent durability
19	Ultra-high-strength HPC	Flowing ultra-high-strength concrete	180–239	≥150	Can be pumped, vibration required, excellent durability
20		High flowing, ultra-high-strength concrete	≥240	≥150	Can be pumped, vibration required, excellent durability

In the UK, Imperial Chemical Industries (ICI) and Oxford University successfully developed a macro defect free (MDF) cement whereby its compressive strength reached 300 MPa, its flexural strength reached 150 to 200 MPa, and a modulus of elasticity of 50 GPa was obtained. The cement was prepared by adding water-soluble polyacrylamide into the cement with a very low water/cement ratio (0.10 to 0.15) and by mixing with a high speed shear mixer so that it could be extruded. In the US, Della M. Roy made a cement paste with a compressive strength of 655 MPa from ordinary cement with a water/cement ratio of 0.1 heated under a temperature of 250°C and a pressure of 50,000 psi (1 psi = 4894.8 Pa). At present, active powder concrete that can reach a compressive strength of 200 to 800 MPa is being widely studied throughout the world. The composition of this concrete includes, apart from the coarse aggregate, additions of silica fume, steel fiber, quartz sand, and even micro steel balls as an aggregate, with low water/cement ratio; it is formed by thermo-pressing. These studies reveal the potential of the strength of cement paste and concrete. However, due to expensive raw materials and the complicated technology used, there is a long way to go before this application is used in civil engineering. But the idea of eliminating these defects can serve as a guide in preparing high strength or super-high-strength concrete.

The technical approaches for preparing super-high-strength high performance concrete discussed in this book are based on ordinary materials and general technology for the convenience of its use in the future. Thus, the technical approaches used are as follows.

2.3.1 Portland cement + active mineral additives + superplasticizer

This approach utilizes the general technology used throughout the world to prepare high strength concrete. In ordinary concrete, to ensure there is sufficient workability of the concrete mix during construction, the water dosage is much greater than that required to fully hydrate the cement. Usually, the water required for hydration of the cement is around 15% to 20% of its mass. However, in practice, the water dosage is closer to 50% to 70% of its mass or even more. Residual water evaporates after hardening of the cement leaving a large quantity of pores of various diameters in the cement paste and at the cement paste–aggregate interface. There are also capillaries and microcracks due to bleeding, shrinkage of concrete, and thermal stress. Such defects are the main causes for reduction of strength and other poor performance. Consequently, diminishing or reducing such defects as much as possible to improve the structure of the concrete is required to produce 100- to 150-MPa super-high-strength high performance concrete. The key is the addition of superplasticizer, to reduce water dosage as much as possible, but also to ensure the necessary fluidity of the concrete mix.

To improve the phase composition of hydrates and their quality in the cement paste is another requirement to produce 100- to 150-MPa super-high-strength high performance concrete. As is well known, there are various cementitious materials such as calcium silicates, calcium aluminates, calcium sulphoaluminates, and calcium ferrite hydrates, which cement the solid particles in concrete into a whole. Among these hydrates, the most abundant and most important is calcium silicate hydrate (C-S-H). There are two different kinds of calcium silicate hydrate: highly basic calcium silicate hydrate and a lowly basic one. The former has low strength and the latter has high strength (see Section 3.1). During hydration of the cement, there is usually a large amount of the highly basic calcium silicate hydrate formed as well as $Ca(OH)_2$ with very low strength and poor stability, so the strength of cement paste from such phase components would not be very high. Therefore, another key in preparing super-high-strength concrete is to reduce the content of the high Ca/Si ratio C-S-H and increase the amount of the low basicity one, as well as reducing or eliminating the $Ca(OH)_2$. The way to do that is to add superfine mineral additives such as silica fume, ground slag, and ground fly ash. In these mineral additives, there is a large amount of active SiO_2, which undergoes a pozzolanic reaction (secondary reaction) with $Ca(OH)_2$ and high Ca/Si C-S-H to form low Ca/Si C-S-H, leading to an improvement in the binder quality in concrete and an increase in its quantity.[106] In such a way, the concrete strength could be increased significantly and super-high-strength high performance concrete prepared.

2.3.2 Ground slag + alkaline component

For this approach, super-high-strength concrete can be prepared from an alkaline–slag binder and its strength can reach 160 MPa.[15] Granulated blast furnace slag is basically a glass in the CaO-Al_2O_3-SiO_2 system. Due to the relatively low CaO/SiO_2 ratio in slag, the hydraulic activity of ground slag is low, and it provides very low strength when simply mixed with water. When activated by lime, gypsum, or Portland cement, slag provides a certain strength, and high strength and super-high-strength concrete cannot be produced from it. However, when activated by compounds from Group 1 of the periodic table (Li, Na, K), the slag shows excellent hydraulic activity. Because alkali hydroxides dissolve rapidly in water and form a great amount of OH^- ions with a strong ionic force, the structure of the slag glass is decomposed rapidly and hydration occurs producing a large amount of low Ca/Si ratio calcium silicate hydrate and aluminosilicate hydrate of alkaline metal, and hardening cement paste is formed. According to experiments carried out by the author, this approach can produce a super-rapid-hardening (R_1 up to 70 MPa), super-high-strength ($R_{28} \geq 100$ MPa), high impermeability (impermeability >4 MPa), and high frost and corrosion-resistant concrete.

2.3.3 Portland cement + superplasticizer + ground sand + autoclave curing

This approach is suitable for applications at a concrete product plant to produce super-high-strength concrete elements, for example, super-high-strength concrete piles, electrical poles, sleepers, bridge spans, and products for special purposes. Though sand is a crystalline SiO_2, after grinding and autoclave curing, it will dissolve rapidly and react with free lime and high basicity calcium silicate hydrate during the hydration of the cement, forming low basicity calcium silicate hydrate. The principle of this approach is similar to that of the first approach, the difference being that instead of an active mineral additive, ground sand is used and instead of normal curing, autoclave curing is applied. The advantage of this approach is that the resources are abundant and the cost of sand is low.

2.3.4 High quality lime + superplasticizer + ground sand + autoclave curing

This is similar to the third approach, but the super-high-strength concrete is produced from high quality lime instead of Portland cement; it is suitable in this case for reducing the use of the cement, but it is not suitable for *in-situ* work.

In this book, only the first approach with cement and a mineral additive for preparing super-high-strength high performance concrete is described.

2.4 THE MECHANISM AND MAIN EFFECT OF WATER REDUCTION WITH SUPERPLASTICIZER

At present the water reduction rate of a high quality superplasticizer is in the range of 25% to 35%, allowing the water/cement ratio of concrete to be lowered to 0.20 to 0.30. With the increased fluidity and improved workability of the mix, the compactness and strength of the concrete can be improved. Therefore, the invention and development of superplasticizers has promoted the progress of concrete technology. As discussed above, in high strength and super-high-strength concrete, an active mineral additive, which has a large specific surface, is used, and without the addition of a superplasticizer, it would absorb a large amount of water. At the same low dosage of water, the mix is harsh, cannot be mixed, and, worse than that, without the addition of mineral admixtures; on the contrary, with the joint action of superplasticizer and these superfine mineral powders, the fluidity of the mix is increased (see Section 3.4). This demonstrates the role superplasticizers play in modern technology to produce high strength and super-high-strength high performance concrete.

The mechanism of water in reducing the effect of plasticizers is due to their surface activity. A surface-active agent is an organic compound, in which there are hydrophilic and hydrophobic moieties. After its addition to water, the surface tension (water–gas interface) and interface tension (water–solid interface) of the water are reduced.

Usually, the hydrophobic end of the surface-active agent is a hydrocarbon while the hydrophilic one is a salt capable of dissociating into various ions, such as $R\text{-}SO_3Na \rightarrow R\text{-}SO_3^- + Na^+$, so the hydrophilic end carries a negative charge. This is an anionic surface-active agent. In addition, there are cationic surface-active agents where the hydrophilic salt dissociates an anion making the hydrophilic end carry a positive charge. Zwitter ion surface-active agents, which dissociate into negative and positive ions, have two hydrophilic radicals, while in non-ionic surface-active agents, the hydrophilic radical does not dissociate ions but has a polar group, such as OH, which absorbs the water molecules and plays the role of hydrophilic radical.

After addition of water, the surface of the cement particle is wetted, and the more it is wetted, the less the water demand is for the same workability (fluidity). The wetting degree of the liquid on the solid surface can be expressed as a wetting angle θ, as shown in Figure 2.1.

As can be seen from Figure 2.1, there are three kinds of interfacial tension on the solid phase of the cement particle, namely, σ_{GL}, the interfacial tension between gas and liquid phases; σ_{LS}, the interfacial tension between liquid and solid phases; and σ_{GS}, the interfacial tension between gas and the solid phases. The critical wetting angle, θ, is given by:

$$\cos\theta = \frac{\sigma_{GS} - \sigma_{LS}}{\sigma_{GL}} \tag{2.1}$$

It can be seen from this equation that when σ_{LS} and σ_{GL} diminish, cosθ increases, so θ reduces and the wetting will be better. After addition of a water-reducing agent, the interfacial tension of the water and that between water and the cement particle reduce, the cement particle is liable to be wetted, and the water dosage can be reduced.

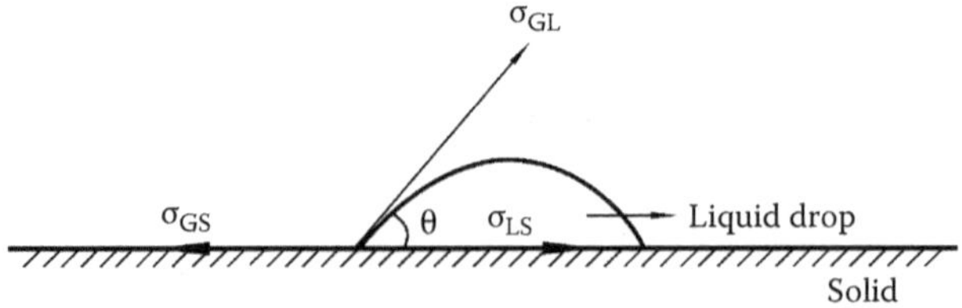

Figure 2.1 Schematic illustration of wetting angle and surface tension.

Figure 2.2 Flocculated structure of the cement paste.

After the cement powder is mixed with water, a flocculated structure will be formed (Figure 2.2), inside of which free water is held or trapped. Such a structure is formed through the association of the solvated water film of the cement particle and agglomeration of similar particles with opposite electric charge as a result of cement mineral hydration. Because a lot of the free water is bonded within the flocculated structure, the workability of the concrete mix is reduced. In a fresh concrete mix without a water-reducing agent, water is normally added to increase the workability of the mix. Consequently, this leads to lowering of its strength, durability, and other properties. If the water bonded in the flocculated structure could be released and become available, workability could be increased without addition of further water. The water-reducing agent can give such an effect.

In a mix to which a water-reducing agent is added, the hydrophobic radical of the adsorbed water-reducing agent is orientated on the surface of the cement particle with the hydrophilic end pointing into the water forming an ordered adsorbed film of water. Due to such orientated adsorption, the whole surface of the cement particle carries the same charge, repelling the cement particles which separate (Figure 2.3); the polar water molecules are absorbed on the surface of the hydrophilic radical, the solvated layer around the cement particle is thickened, and the sliding ability between particles increased. Due to the separation and movement of the cement particles,

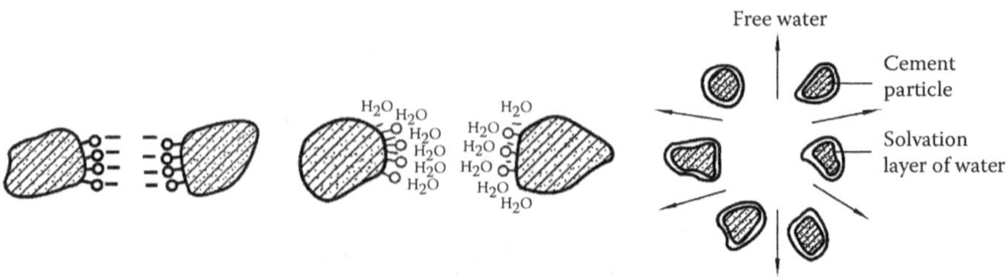

Figure 2.3 Mechanism of the action of water-reducing agents.

the flocculated structure is no longer formed or it is broken up, releasing the bonded water and increasing the workability of the concrete mix. In other words, by keeping the fluidity unchanged, the addition of a water-reducing agent and the water dosage of concrete can be reduced. This is one explanation for the mechanism of reduced water dosage by a water-reducing agent.

A second explanation is the theory of diffusion double layer,[10] where after addition of water to the cement, the reaction of C_3A and C_3S at the surface of the cement particle with water releases cations such as alkalis and Ca^{2+} which diffuse faster than anions such as $H_2SiO_4^{2-}$ and $AlOH_4^-$, so on the surface of the cement particle, the concentration of anions is higher than that of cations forming a negatively charged layer. Ca^{2+} and other cations are adsorbed onto the surface of the cement particle to form an adsorbed layer, its thickness being δ. In addition, the ions at the surface of cement particles apart from being adsorbed have thermal energy and diffuse into the water, forming a diffusion layer; thus the adsorbed and diffusing layer is called a *double layer*. The absorbed layer and the core particle together form a colloidal agglomerate which moves together. A potential exists between the diffusion layer and adsorbed layer that is called *electromotive* or *ζ potential*, as shown in Figure 2.4, where E is thermodynamic potential independent from the ζ potential.

The magnitudes of E and ζ are related to the concentration of the ions in solution. As hydration of the cement proceeds, the cations in the absorbed layer increase, but the cations in the diffusion layer decrease; when the amount of cations has increased so the thickness of the diffusing layer equals zero, the cations in the adsorbed layer neutralize the negative charge on the surface of the colloidal agglomerate, the ζ potential equals zero (Figure 2.5), and the cement paste loses its fluidity.

At present, many superplasticizers are still anionic surface-active agents: after adding the superplasticizer into the cement, the anion decomposed in water solution enters into the absorbing layer and increases the concentration of anions in the absorbing layer, raising the ζ potential and increasing the thickness of the diffusion double layer leading to improved fluidity of the cement paste. This is another interpretation for the mechanism of water reduction.

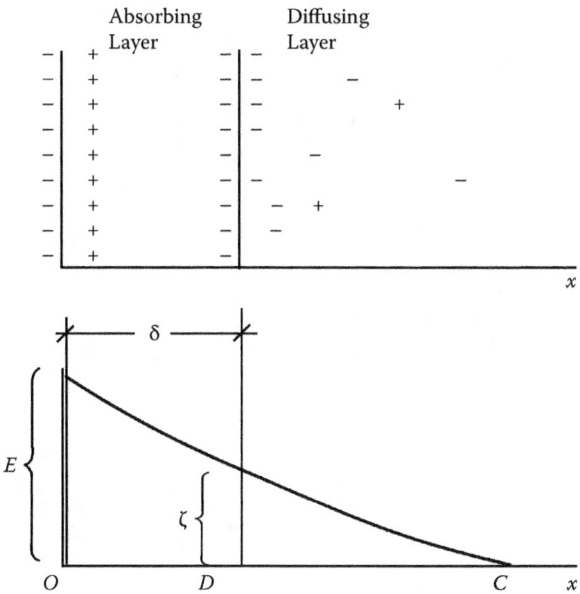

Figure 2.4 Diffusion double layer.

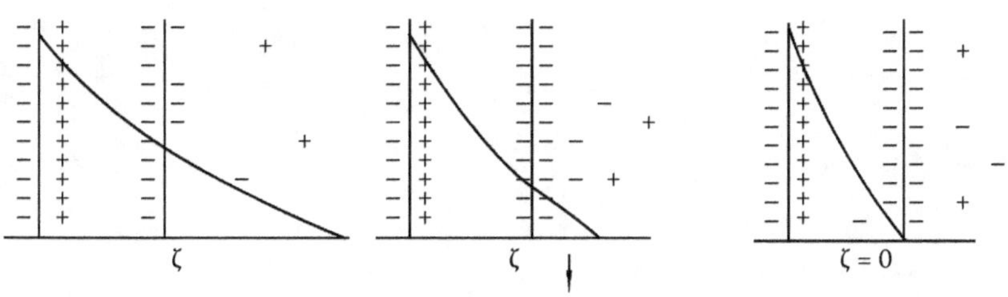

Figure 2.5 Decrease of ζ potential.

Chapter 3

Important effects of active mineral admixtures

3.1 STRENGTH OF CALCIUM SILICATE HYDRATES AND OTHER HYDRATION PRODUCTS

Concrete is an artificial stone formed by cementing coarse and fine aggregates with calcium silicate hydrates and other hydration products formed from the hydration of Portland cement. Therefore, it is worth discussing the strength and other performance characteristics of the various calcium silicate hydrates.

Ilyushin et al.[16] investigated the tensile strength of a crystalline whisker of a synthetic calcium silicate hydrate. The result showed that the tensile strength of this whisker synthesized as a low Ca/Si ratio calcium silicate hydrate (C/S <1.5) and reached 1300 MPa, but one with a higher ratio had only half the strength of the former.[16] Timashev[17] synthesized single crystals of the low Ca/Si ratio calcium silicate hydrates tobermorite, foshagite, and xonotlite which were shown to have tensile strengths around 1300 to 2000 MPa, but that of high Ca/Si ratio calcium silicate hydrate is only 700 to 830 MPa.

In the early 1960s, Kuatbaev[18] synthesized small samples of certain types of pure calcium silicate, calcium aluminate, and calcium silicoaluminate hydrates and measured their bending strengths (Table 3.1). It can be seen from Table 3.1 that CSH(B), tobermorite, and xonotlite as low Ca/Si calcium silicate hydrates have high bending strengths, but $C_2SH(A)$ and $C_2SH(C)$, as high Ca/Si calcium silicate hydrates, have low bending strengths. These data coincide with the test results given in the literature[16,17] on the tensile strength of single crystals.

Why do the low Ca/Si ratio calcium silicate hydrates give such high strengths? Ilyushin et al.[16] suggested that in low Ca/Si calcium silicate hydrates, the degree of polycondensation of the silicon–oxygen chain was much higher, but in high Ca/Si calcium silicate hydrates, there is a lower degree of polycondensation. The study by Gu Zhangzhao[19] showed that the high sulfur calcium sulfoaluminate, ettringite, had a high compressive strength (13.07 MPa), which was decreased greatly after carbonation

Table 3.1 The flexural strength of calcium silicate hydrates in MPa

Name of hydrate[a]	*After synthesis*	*After carbonation*	*After drying and wetting cycle*	*12 months in 5% Na_2SO_4 solution*	*12 months in 2.5% $MgSO_4$ solution*
Tobermorite	3.5	3.0	2.3	3.1	0.9
CSH(B) [CSH(I)]	4.0	3.3	2.6	3.6	0.7
Xonotlite	8.3	7.5	6.0	6.9	1.6
$C_2SH(A)$ [α-C_2S hydrate]	0.5	5.0	1.6	0.4	0.5
$C_2SH(C)$ [γ-C_2S hydrate]	0.8	2.8	1.4	0.6	0.7
C_3AH_6	2.4	3.2	2.6	Broken after 10 d	Broken after 15 d
C_3ASH_4	1.9	2.7	2.3	Broken after 45 d	Broken after 45 d

[a] Bogg's symbols and Taylor's nomenclature in brackets.

(5.97 MPa), but the low sulfur calcium sulfoaluminate (monosulfate) had a low compressive strength (6.15 MPa), which after carbonation increased (10.93 MPa).

The strength of the crystal changes, not only within its structure, but in its shape and dimensions as well. The smaller the dimension of the crystal, the higher its strength will be. This is due to the fact that the smaller the dimension of the crystal, the better its integrity is with fewer defects such as dislocations, pores, or cracks. Under loading, stress is concentrated at these defects from where the damage initiates. Therefore, the measured strength of the crystal is far less than its theoretical one.

Butt et al.[19] suggested that from the viewpoint of crystal intergrowth, crystals of low Ca/Si calcium silicate hydrate have small sizes and large specific areas so the crystal intergrowth from such crystals makes a large number of contact points, giving rise to high strength. The crystal intergrowth from the coarse crystals of high Ca/Si calcium silicate hydrate creates few contact points, so the strength is low. The size and specific areas of calcium silicate hydrate crystals are given in Table 3.2.

Ilyushin[16] pointed out that the strength of cement paste was influenced not only by the strength of mono-crystals but also by the connection of the microskeleton formed by the single crystals, so the formation of double crystals from calcium silicate hydrate is a special case. In cement pastes, the basic patterns form various regular crystal intergrowths, where the formation of various intergrowth crystals may be the result of double crystallization or may be accidental. Microhardness determination in the contact zone of intergrowth showed that 85% of the hardness measured was equal to or more than the microhardness of the crystals forming the intergrowth.

Table 3.2 The size and specific area of the calcium silicate hydrate crystals

Name of hydrate[a]	*Shape*	*Size (μm)* Length × (width)	*Size (μm)* Thickness	*Specific surface area ($m^2.g^{-1}$)*
CSH(B) [CSH(I)]	Fiber	<1	0.0084–0.011	80–100
Tobermorite	Flake	<2	0.013	60
Xonotlite	Fiber	6–10	0.1	18
$C_2SH(A)$ [α-C_2S hydrate]	Prism, flake	100 × 40	0.27	2.6
$C_2SH(B)$ [β-C_2S hydrate]	Fiber, needle	<30	0.25	6.8
$C_2SH(C)$ [γ-C_2S hydrate]	Grain	Φ1	—	2.6
C_3SH_3	Fiber, needle	10–30	0.01–0.03	74.5

Source: Chongqing Institute of Architecture and Engineering and Nanjing Institute of Technology, *Science of Concretes*, Beijing: China Building Industry Press, 1983 (in Chinese).

[a] Bogg's symbols and Taylor's nomenclature in brackets.

Within the contact zone of crossed double crystals, the measured hardness was 4600 MPa; for the contact zone of ordinary double crystal, 3500 MPa, and within the contact zone of intergrowth, 3800 MPa, indicating the high structural homogeneity of the contact zone of calcium silicate hydrates. Accordingly, the number of faults such as cracks, voids, and dislocations in the contact zone was small.[16] This basic research on the strength of calcium silicate hydrates laid the foundation for our preparation of super-high-strength cement paste and concrete.

Although in a real cement paste found in concrete, the cementitious material is not composed from a single hydrate, but a composition of several hydrates, it is possible to obtain a cementitious material with particular hydrates as its main component. Yambor[20] has studied the strength of a cementitious material specimen from a composition of several hydrated minerals (Figure 3.1) and showed that a specimen from the composition of tobermorite + CSH (B) has the highest strength. Taking this strength as 100%, the next strongest was the CSH (B) specimen or a specimen from CSH (B) + C_2SH_2 (relative strength 56% to 62%); then follows the composition from gehlenite hydrate (70% to 80%) + CSH (B) (20% to 30%) (relative strength 28% to 38%); the next is the composition of hydrogarnet (70% to 80%) + CSH (B) (20% to 30%) and the lowest is the composition of C_3AH_6 + hydrogarnet (the relative strength is 3% to 4%).

The stability of the low Ca/Si ratio calcium silicate hydrates is higher than the high Ca/Si ratio ones (Table 3.3); for example, the solubility of low Ca/Si calcium silicate hydrate CSH (I) in water is low (0.05 kg/m^3) while that of high Ca/Si calcium silicate hydrate, C_2SH_2, is more than 20 times higher (1.4 kg/m^3).

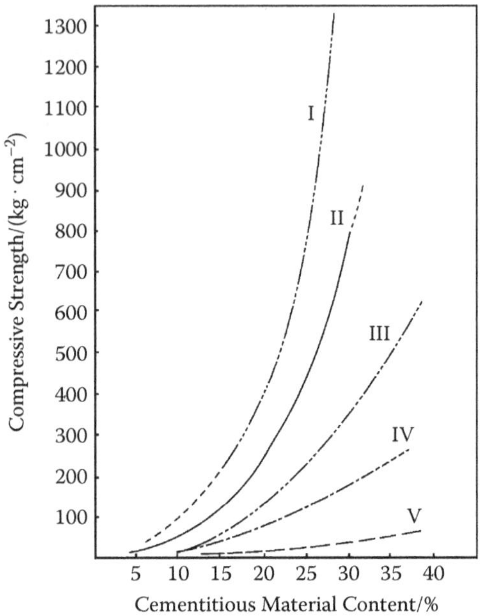

Figure 3.1 Relationship between type and content of cementitious material with the strength of the sample. I = Tobermorite + CSH (B); II = CSH (B) or C_2SH_2 + CSH (B); III = gehlenite + CSH (B); IV = hydrogarnet + CSH(B); V = C_3AH_6 + hydrogarnet. (From Y. Yambor, Discussion on the 5th International Congress of Cement Chemistry, in the Report on the 5th International Congress of Cement Chemistry, Stroiizdat, 1973 [in Russian].)

In the hydrates formed from Portland cement, there is another important hydrate, free lime or portlandite, which makes up 20% to 25% of the total hydrates. Portlandite has low strength with very poor stability and is the first substance to be attacked in a corrosive environment. It is often present as coarse crystals separated from the remaining hydrated cement paste and oriented at the interface between the cement paste and the aggregate leading to a weakening of the interfacial bond between the cement paste and aggregate.

The study of the strength of the various hydrates is inadequate and not systematic and further investigation is needed. However, the existent research results have shown some characteristics of the strength of these hydrates. The author thinks that the strength characteristics of calcium silicate hydrates and other hydrates show us a way to prepare super-high-strength cement paste and concrete. The main purpose of this work is to transform the high Ca/Si ratio hydrated cementitious material and free lime in the concrete into a cementitious material with a low Ca/Si ratio, calcium silicate hydrate, as its main component.

Table 3.3 Solubility of hydrates and aggregate in concrete

Item	*Hydrate and mineral*	*Crystal formula*	*Chemical formula*	*Lattice energy ($kJ.mol^{-1}$)*	*Solubility (kg/m^3)*
Portland cement concrete	Calcium hydroxide		$Ca(OH)_2$	—	1.3
	C_2SH_2		$2CaO\cdot SiO_2\cdot nH_2O$	13218.4	1.4
	C-S-H (I)		$5CaO\cdot 6SiO_2\cdot nH_2O$	—	0.05
	Tetracalcium aluminate tridechydrate		$4CaO\cdot Al_2O_3\cdot 13H_2O$	—	1.08
	Tricalcium aluminate hexahydrate		$3CaO\cdot Al_2O_3\cdot 6H_2O$	15158.4	0.56
	Calcium sulphoaluminate hydrate		$3CaO\cdot Al_2O_3\cdot 3CaSO_4\cdot 32H_2O$	—	High
Alkaline-slag cement concrete	C-S-H (I)		$5CaO\cdot 6SiO_2\cdot nH_2O$	—	0.05
	Xonotlite	$Ca_6[Si_6O_{17}](OH)_2$	$6CaO\cdot 6SiO_2\cdot H_2O$	17420.0	0.04
	Gyrolite	$Ca_4[Si_6O_{15}](OH)_2.4H_2O$	$2CaO\cdot 3SiO_2\cdot 2.5H_2O$	17554.5	0.05
	Hydrogarnet		$3CaO\cdot Al_2O_3\cdot 1.5SiO_2\cdot 3H_2O$	19024.4	0.02
	Thomsonite	$NaCa_2[Al_5Si_6O_{20}].6H_2O$	$Na_2O\cdot 4CaO\cdot 5Al_2O_3\cdot 10SiO_2\cdot 12H_2O$	17142.2	0.05
	Hydronepheline		$Na_2O\cdot Al_2O_3\cdot 2SiO_2\cdot 2H_2O$	19895.4	0.02
	Natrolite	$Na_2[Al_2Si_3O_{10}]\cdot 2H_2O$	$Na_2O\cdot Al_2O_3\cdot 3SiO_2\cdot 2H_2O$	20481.5	0.02
	Analcite	$Na[Si_2AlO_6]\cdot H_2O$	$Na_2O\cdot Al_2O_3\cdot 4SiO_2\cdot 2H_2O$	21214.4	0.02
Aggregate	Calcite		$CaCO_3$		0.014

Sources: P.V. Krivenko, *Durability of Alkali-Slag Concrete*, Budivelnik, 1993 (in Russian) and Pu Xincheng, On the Extra Durability of the Concrete Project. *Concrete*, 2000 (1) (in Chinese).

3.2 THE POZZOLANIC REACTION AND STRENGTHENING EFFECT OF ACTIVE MINERAL ADMIXTURES

The active mineral admixture is a substance that contains large amounts of amorphous SiO_2 and Al_2O_3. Many are natural materials such as pozzolanic ash, zeolites, diatomaceous earth, pumice, and so forth that are referred to as pozzolanic materials. The active mineral admixtures used in modern concrete, however, are mostly industrial wastes or by-products such as fly ash, blast furnace slag, phosphorous slag, silica fume, and so forth. Before their application, such materials should be examined for their quality and whether they have the necessary fineness. Much of such material could serve as active mineral admixture for concrete only after grinding.

The basic minerals in Portland cement clinker are C_3S, C_2S, C_3A, and C_4AF. C_3S and C_2S make up about 75% of the cement mass and contribute the bulk of the strength on hydration. The hydrates formed are high CaO/SiO_2 calcium silicate hydrates with a Ca/Si ratio of 1.6 to 2.0 along with $Ca(OH)_2$ in a reaction that is approximately

$$2\ (3CaO.SiO_2) + 6H_2O \rightarrow 3\ CaO.2SiO_2.3H_2O + 3Ca(OH)_2$$

$$2\ (2CaO.SiO_2) + 4H_2O \rightarrow 3\ CaO\cdot 2SiO_2.3H_2O + Ca(OH)_2$$

Addition of an active mineral admixture can improve the composition of the cementitious material in a cement paste; in particular, it will reduce or eliminate the $Ca(OH)_2$ because the SiO_2 in the active mineral admixture undergoes a pozzolanic reaction with $Ca(OH)_2$ and the high Ca/Si ratio calcium silicate hydrates to form a lower Ca/Si ratio calcium silicate hydrate with higher strength and better stability.

$$(0.8\text{–}1.5)Ca\ (OH)_2 + SiO_2 + \{n - (0.8\text{–}1.5)\}\ H_2O \rightarrow (0.8\text{–}1.5)\ CaO\cdot SiO_2\cdot nH_2O$$

$$(1.5\text{–}2.0)\ CaO\cdot SiO_2\cdot nH_2O + xSiO_2 + yH_2O \rightarrow ((0.8\text{–}1.5)\ CaO\cdot SiO_2\cdot qH_2O)$$

Thus, the hydration product obtained after the pozzolanic reaction of the active mineral admixture is a low Ca/Si ratio calcium silicate hydrate, improving the interface structure between the cement paste and aggregate and raising the strength of the concrete significantly. The pozzolanic reaction proceeds slowly in the hardened concrete gradually strengthening the concrete; consequently, the pozzolanic reaction is called a secondary reaction.

3.3 THE FILLING EFFECT OF ACTIVE MINERAL ADMIXTURE

In studying the materials used for concrete, a significant emphasis is placed on the grading of coarse and fine aggregate particles to attain an optimum filling in order to reduce the porosity of the hardened concrete. However, the grading of the cementitious materials (cement + additive) is usually neglected. Generally, the average particle size of the cement is 20 to 30 μm, giving a deficiency of particles less than 10 μm so filling of the pore space between the cement particles is not ideal. To improve the filled state of the cementitious material, superfine particles should be added to the cement to give a better composition of coarse and fine particles.

Nayataki[22] studied the filling character of coarse and fine particles. Suppose the porosity of both coarse particle D_1 and fine particle D_2 are 0.4. After combining these two kinds of particles, the porosity could reach a low value, as shown in Figure 3.2. When the volume percentage of the coarse particles $V_1/(V_1 + V_2)$ is 0.7, decreasing the particle size ratio D_2/D_1 gives rise to decreasing porosity and an increasing filling rate. When the ratio is 0.5, the increase is not significant, but for the ratios 1/5, 1/10, and 1/100, the ratio increases greatly.

If a superfine mineral admixture such as superfine fly ash or superfine slag (with an average particle size of 3 to 6 μm) is added into the cement, the filling character of the particles of the cementitious material can be improved greatly, enhancing the compactness and impermeability of the hardened cement paste giving an increased strength due merely to an improvement in the filling character of cement particles. If a reasonable amount of silica fume is added to the blend, its smaller particle size (average particle size 0.10 to 0.26 μm) is one order smaller than that of superfine slag or superfine fly ash, and two orders smaller than that of cement, so the

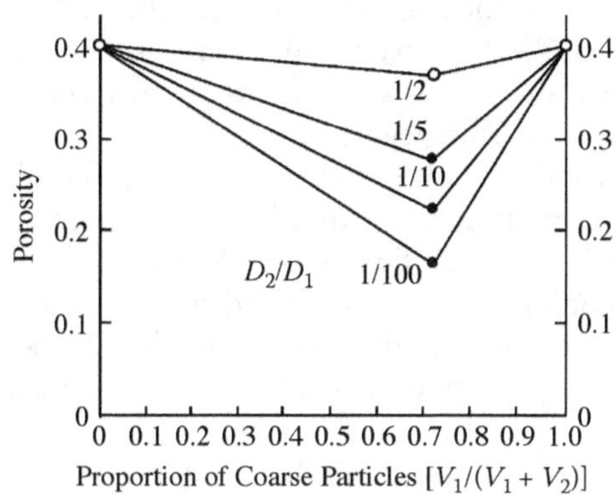

Figure 3.2 Improvement of filling rate by composition of different particle sizes.

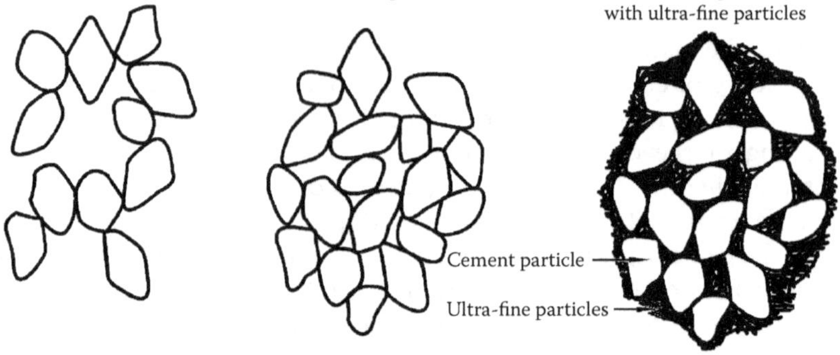

Figure 3.3 Sketch of filling effect of superfine particles in cement paste.

silica fume particles fill the space between superfine slag or superfine fly ash particles, further enhancing the compaction of the cementitious material (Figure 3.3) and increasing the strength. Certainly, adding silica fume alone gives a good filling effect, increasing the compactness of the concrete and improving durability.

3.4 THE PLASTICIZING EFFECT OF ACTIVE MINERAL ADMIXTURES

Not only does the active mineral admixture provide a good filling effect with increased strength, but some mineral admixtures such as silica fume and fly ash may also provide an excellent plasticizing effect in cement pastes, mortar, or concrete. Table 3.4 gives the improvement in fluidity of concrete mix for three commonly used mineral admixtures: fly ash, slag, and silica fume. It can be seen that for concretes without the additives, the fluidity of their mixes was small—some even had a zero slump. With additives, the fluidity increased, in some cases significantly before the increase of fluidity slowed down or even became negative.

Little is known about the plasticizing effect of mineral admixtures such as silica fume. It seems that such an additive with its very small particle size (0.1 μm order), large specific surface (15 to 25 m^2/g), and high water demand would not enhance but reduce fluidity. In fact, during mixing with a superplasticizer, the fine round particles of silica fume are covered by a layer of surface-active compound, so between cement and additive particles, there is electric repulsive force between the particles. As the round particles of silica fume are much smaller than those of cement, they play the role of *ball bearing* between the cement particles, leading to increased fluidity of the cement paste. What is more important is that in the case of a

Table 3.4 Influence of fly ash, slag, and silica fume on fluidity of concrete mix

Content of cementitious material (kg/m³)	*W/C*	*Active mineral admixture*		*Cement (%)*	*Slump (mm)*	*Compressive strength (MPa)*		*Reference*
		Name	*Dosage (%)*			*3 d*	*28 d*	
600	0.265	Fly ash	0	100	42	53	73.2	Ma Jianxin et al. [a]
			9	91	102	53.1	76.7	
			17	83	106	49.6	78.3	
			23	77	205	43.5	76.3	
			30	70	190	40.2	70.1	
583	0.30	Slag	0	100	30		93.0	Zhang Xiong et al. [b]
			20	80	120		86.0	
			30	70	160		92.0	
			40	60	170		96.0	
			50	50	230		91.0	
			60	40	230		78.0	
650	0.20	Silica fume	0	100	0	85.8	108.1	Pu Xincheng et al. [c]
			5	95	17	89.7	126.8	
			10	90	235	99.3	140.7	
			15	85	240	92.3	136.9	

Note: In the mix proportion of each group of concretes, the same water-reducing agent with the same dosage was added.

[a] Ma Jianxin et al., Preparation of Plastic Highly-Flow High Strength Concrete with Water Reducing Type of Fly Ash and Slump Loss Control, *Concrete*, 1997 (1) (in Chinese).

[b] Zhang Xiong et al., Characteristics and Mechanism of Action of the Composite Material from High Strength Concrete and Slag, *Concrete and Cement Products*, 1997 (3) (in Chinese).

[c] Pu Xincheng, Yan Wunan, Wang Cong et al., Contribution of the Silica Fume to the Strength and Fluidity of the 150 MPa Super High Strength High-Flowing Concrete, *Concrete and Cement Products*, 2000 (1) (in Chinese).

paste without silica fume or other mineral admixtures, the pores between the cement particles are not filled by solid particles, less water is on the surface of the cement particles, and more water fills the pores between the cement particles. When the pores between the cement particles are filled by fine grains of silica fume or an active mineral admixture substituting for water in the pores, the thickness of the water layer separating the particles is greater and the fluidity of the mix increases. In addition, the relative density of active mineral admixtures (2.29 to 2.26 for silica fume, 2.42 for superfine fly ash, and 2.92 for superfine slag) is less than the relative density of the cement (3.1), so the admixture volume is larger than that of the paste from the substituted cement, another reason for the increased fluidity.

Shi Yunxing et al.[26] suggested that the surface energy of powder from glassy material is higher than that from non-glassy material. For example, the zeta potential of colloid from fine slag ($S = 5200\ cm^2/g$) in water with the addition of 0.9% superplasticizer NF and a water/powder ratio of 0.5 is –4.9 mV, whereas the zeta potential for cement colloid is –1.7 mV, while that for ground non-glassy river sand ($S = 5200\ cm^2/g$) is –1.1 ~ –0.5 mV. As the repulsive force between the particles is directly proportional to the square of zeta potential, the surface energy of slag glass powder is large and more plasticizer could be absorbed. There would be more electric repulsive force of its colloid, and therefore more dispersing effect between the particles. This gives the filling effect of the powder full play, in turn leading to a large increase in the fluidity of the paste.

The author suggests that to allow the plasticizing effect of active mineral admixture to have full effect, the action of the plasticizer should coordinate with the effect of the additive. For example, in Shi Yunxing et al.[26] (Figure 2) the plasticizing effect of silica fume is not obvious, because the dosage of the plasticizer NF is insufficient for the cement paste with silica fume.

3.5 HEAT OF HYDRATION AND THE EFFECT ACTIVE MINERAL ADMIXTURE HAS ON REDUCING THE MAXIMUM RATE OF HEAT OUTPUT

In high strength and super-high-strength concretes, the increased cement content gives rise to increased temperature of the concrete caused by hydration, so in large-scale structural elements, the inner temperature could reach more than 60 to 70°C in a short time.[27] This large difference in temperature between the inside and outside of the concrete can lead to temperature-induced cracking of the concrete. Therefore, it is important to study the hydration heat of super-high-strength concrete.

In concrete, the heat of hydration depends on the type of cement, cement content, water/binder ratio (water/cement ratio), and so forth. It is also well known that cement with a higher mass proportion of C_3A and C_3S will give more hydration heat. For certain types of cement, concrete with more cement content will naturally evolve more total heat.

High strength concrete (the water/binder ratio is usually 0.3 to 0.4) and super-high-strength concrete (the water/binder ratio is usually 0.2 to 0.3) were obtained on the basis of lowering the water/binder ratio, so how does the water/binder ratio affect the heat of hydration? Many experts have come to the conclusion through research that with a decrease of water/binder ratio, the heat of hydration was lowered,[10,13] as shown in Table 3.5 and Figure 3.4.

It can be seen from Table 3.5 and Figure 3.4 that with the reduction of the water/binder ratio, the total heat of hydration per kilogram cement or binder was lowered significantly. For example, when the water/cement ratio was lowered from 0.68 to 0.26, the hydration heat was reduced by 45.6%, and in the case of a silica fume addition, by 36.5%.

When the water/binder ratio was lowered, a significant reduction in heat of hydration was due to the deficiency of water in the concrete; the degree

Table 3.5 Influence of water/binder ratio (water/cement ratio) on the heat evolution of concrete (20°C, 7 days)

Water/binder ratio (water/cement ratio)	*Heat evolution per kg cement for concrete without silica (fume/k)*	*Heat evolution per kg binder for silica fume (concrete/k)*
0.68	454	356
0.40	333	295
0.31	307	271
0.26	247	226

Source: Odd E, Gjorv, High Strength Concrete, in *Advances in Concrete Technology*, CANMENT, 1992.

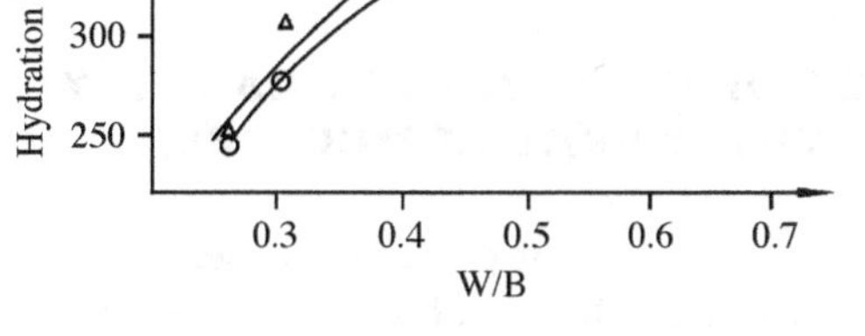

Figure 3.4 Relationship between hydration heat of the cement and water/binder ratio. (From Odd E., Gjorv, High Strength Concrete, in *Advances in Concrete Technology*, CANMENT, 1992.)

of hydration of the cement was diminished. Besides, it can be seen from Figure 3.4 that the addition of silica fume reduced the hydration heat of the cementitious material. Therefore, in high strength and super-high-strength concretes, addition of active mineral admixtures will lower the hydration heat and reduce the temperature rise of the concrete, although some researchers[10] suggest that highly active silica fume makes little contribution to decreasing the temperature rise.

At Tsinghua University, the adiabatic temperature rise of super-high-strength, high strength, and ordinary concretes made with addition of silica fume or fly ash and with strengths in the C100 to C35 range has been determined (Table 3.6). It can be seen from the table that with diminishing water/binder ratio, the adiabatic temperature rise slowed down. Adding fly ash (mix No. 3 compared with those of No. 4 and No. 5, with the same water/binder ratio) lowered the temperature rise as well.

Active mineral admixtures, especially those with low pozzolanic activity such as fly ash and zeolite powder, play a significant role in lowering temperature rise. By keeping the total amount of binder in the concrete unchanged, mineral addition reduces the cement content in concrete. Therefore, the heat evolution due to cement hydration in concrete is reduced, and it will be more so with the addition of more additive. Though there is heat of hydration due to the pozzolanic reaction of the active mineral admixture in concrete, such a reaction will lag behind the main hydration of cement and proceed for a longer time. As some of this heat can be gradually lost to the atmosphere, the heat from the secondary reaction makes only a small contribution to the maximum temperature rise of concrete. However, for some active mineral admixtures with more pozzolanic activity such as silica fume, their pozzolanic reaction occurs early lagging less behind the hydration of cement, the secondary reaction is marked and more heat is generated in the secondary reaction. The mineral admixtures have less effect in diminishing the maximum temperature rise, so that in such cases, this cannot be taken into consideration. In construction work, countermeasures should be taken to avoid excess temperature rise or temperature differential between the outside and inside of the concrete, which leads to temperature-created cracks.

3.6 THE EFFECT OF IMPROVING DURABILITY WITH ACTIVE MINERAL ADMIXTURES

As illustrated above, the addition of active mineral admixtures and their secondary reaction with free lime and high basic calcium silicate hydrates to form stronger and more stable low Ca/Si ratio calcium silicate hydrates leads to improvement in the composition of the hydrated binder, and reduces or eliminates free lime, so it will play a significant role in enhancing

Table 3.6 Influence of water/binder ratio, cement content, and mineral admixture on adiabatic temperature rise of concrete[10]

No.	*W/B*	*Grade*	*Cement (kg/m³)*	*Fly ash (kg/m³)*	*Silica fume (kg/m³)*	*Time of rapid temperature rise (h)*	*Time for start of stable temperature rise (h)*	*Time for maximum temperature rise (h)*	*The highest temperature rise (°C)*
1	0.236	C100	500		50	26		31	41.54
2	0.260	C80	450		50	29	42	58	42.16
3	0.290	C60	550			22	27	216	52
4	0.300	C60	450	100		22	24	31	45.88
5	0.300	C50	385	165		22	92	216	44.2
6	0.328	C40	382	168		14	21.5	71.5	50
7	0.365	C35	382	168		24.5	36.5	131	53

Note: Ji-dong 525 OPC (Chinese standard GB 175-1992) was used in mix No. 1 through No. 5; Shou-du 525 OPC was used in mix Nos. 6 and 7.

the durability of concrete. For instance, the resistance of concrete with active mineral admixtures to sulfate corrosion is enhanced due to the lack or absence of free lime in the hardened cement paste, and there is no further ettringite formation, resulting in damage due to its expansion. What is more, in the alkali–aggregate reaction, because of the addition of an active mineral admixture, a large amount of low Ca/Si ratio calcium silicate hydrate will be formed which can absorb and fix large amounts of Na+ and K+ ions, effectively reducing the alkali content in the pore solution, leading to significant reduction in damage caused by the alkali–aggregate reaction.

The active mineral admixtures in high performance concretes are all in the superfine form: they fill the spaces between cement particles or the pores in the interface, making the structure of cement paste and interface more compact, blocking the passage to permeation, enhancing impermeability so it is difficult for water and corrosive media to enter into the concrete. As a consequence, durability can be significantly enhanced.

In a case where water cannot penetrate the concrete, there is little water to freeze. As a result, under the action of a freeze–thaw cycle, the frost resistance of high strength concrete is greatly enhanced. For example, the study of Cao Jianguo, Li Jinyu et al.[28] showed that the number of rapid freezing and thawing cycles for high strength concrete C80 (binder 590, water cementitious ratio 0.26, silica fume 6.8%, slag powder 8.5%) and super-high-strength concrete C100 (binder content of 550 kg/m^3, water/cement ratio of 0.22, and silica fume replacement of 9%) both with active mineral addition could reach 1200 cycles. While the dynamic modulus of elasticity was 92.7% and 96.9%, respectively, the mass loss for both concretes was 0. Of course, such high frost resistance is related to the low water/binder ratio, but is also closely related to the addition of active mineral admixtures. In the case of a C60 concrete with the same water/binder ratio of 0.26 but without the active mineral admixtures, the number of freezing and thawing cycles was only 250.

One negative influence the addition of active mineral admixtures has is the reduction of the basicity of the concrete, so the resistance to carbonation is weakened, and its protective ability for reinforcing steel is lowered. However, the rate at which the basicity of concrete was lowered is not as large as expected. A number of studies on basicity of cement with large additions of fly ash by the author[29] show that in cases of fly ash addition at 0, 30%, 40%, 50%, 60%, and 70%, the pH values are 12.56, 12.50, 12.46, 12.24, 12.15, and 12.06, respectively. This illustrates that even in the case of 70% fly ash addition, the pH value of cement mortar is more than 12, which is higher than the minimum alkalinity 11.50 allowed for reinforced concrete structures. If an additive with high CaO content such as slag is added, the basicity is guaranteed. Besides, the addition of active mineral admixtures improves the compactness and impermeability of high performance concrete, so it is difficult for water, or even O_2 or CO_2, to penetrate the concrete, enhancing the ability of the concrete to protect the reinforcing steel from corrosion.

The author concludes that it is possible to prolong the service life of concrete to 500 to 2000 years using technical measures consisting of the addition of active mineral admixtures and superplasticizers.[3]

3.7 METHODS FOR ADDITION OF ACTIVE MINERAL ADMIXTURES

Active mineral admixtures can be added directly into cement as well as into mortar or concrete. In the case of addition into cement, the clinker and mineral admixture should be ground separately and then mixed in the correct proportions. This is due to the different grindability of clinker and additive. Using the traditional technology of mixed grinding for clinker and slag, the poor grindability of slag means that when the blend is *fully* ground, the fineness of the slag is less than that of the clinker; therefore, the bleeding of slag Portland cement is usually high with lower frost resistance. To be an effective active mineral admixture, extra grinding is required, so that its fineness is much higher than that of the clinker to ensure the full effect of its pozzolanicity.

The addition of active mineral admixture into concrete is flexible. Choices can be made regarding the type of additive and the number of additives used, and the dosage of each additive can be varied arbitrarily. Due to the variation in additive performance, caused by factors such as chemical composition, structural state, amount of active component, and fineness, each kind of additive has its own specific characteristics. For example, when fly ash is used, the early strength of concrete is reduced, but later long-term strengths are well developed. The temperature rise is reduced and the concrete has a strong ability to prevent alkali–aggregate reaction. In the case of large dosages, the basicity reserve of material may not be enough and its resistance to carbonation may be lowered. When slag is added, early strength in concrete develops well, but later strength is less than that of concrete containing fly ash. Due to the high mass proportion of CaO in slag, slag addition can reach 50% to 70%, without a rapid fall in strength and large reduction in basicity. Silica fume has the highest fineness, the largest mass proportion of active SiO_2, and the highest activity index. Consequently, its pozzolanic reaction is fast with initial high early strengths and continued strength development at later ages. With its high fineness, excess or large dosage leads to a rapid loss in fluidity. It is used at lower dosages than that of slag or fly ash, but with low availability and high cost, large dosages will lead to increased cost.

Therefore, different additives have their own advantages and disadvantages, so the choice of additive should be based on their characteristics and the work requirements. A single additive could be used, where the dosage is chosen according to the performance characteristics of the additive; for example, the optimum dosage for silica fume is 10% to 15%, for superfine fly ash 20% to 30%, and for superfine slag 30% to 50%. The combination of several additives (usually two to three types) could be used, where the composition is based on the characteristics of one or more additives to gain a supplementary effect of producing concrete with excellent properties in all aspects.

As shown above, in the case of combined additives, a complex pozzolanic reaction can occur. For example, while the activity index of fly ash or slag[30,31] is lower than that of silica fume, if part of the silica fume is substituted by fly ash or slag, or a combination thereof, the strength of the concrete may be close to or exceed that of concrete with the silica fume alone,[30,32] illustrating the combined effect. It should be considered that such an effect could be a physical one of mutual compaction by filling with additive particles of different sizes.

At present, the two outstanding achievements in modern concrete science are, first, the manufacture and application of superplasticizers and other efficient admixtures, and second, the research, application, and development of active mineral admixtures. The importance of the application of active mineral admixtures is far more than simply economic, saving of cement clinker, and the environmental advantages of utilizing industrial wastes and by-products. There is overall improvement of concrete properties and further development of concrete science. Additives are necessary for developing high strength and super-high-strength concretes giving rise to high durability and super-high-durability concretes, and other functional cement-based materials. However, further research is needed, as the current knowledge of active mineral admixtures is insufficient. New types of mineral admixtures need to be studied and developed, the standards and regulations for their application must be compiled, the advantages advertised, and existing enterprises need to be enlarged and brought up to standard, so there is a long way to go. It will be necessary to set up new enterprises to process active mineral admixtures to go into cement production—ready-mix concrete plants and superplasticizer factories to guarantee a stable supply of high quality, high efficient, active mineral admixtures. Only in this way can high performance concrete be realized and meet the requirement of the rapid growth in China and in the world.

Chapter 4

Analysis of the pozzolanic effect and strength composition

4.1 CLASSIFICATION OF ACTIVE MINERAL ADMIXTURES AND FACTORS THAT INFLUENCE THEIR QUALITY

Basically, there are three types of active mineral admixtures:

1. Additives with potential hydraulic properties, such as granulated blast furnace slag and other granulated slags. The characteristics of these types of additives are a high CaO content (35% to 48%) with a high proportion of active SiO_2 and Al_2O_3. In comparison to cement clinker, where CaO makes up 63% to 65% of the total oxides, the mass proportion of CaO in slag is lower and the slag has little or no hydraulicity when used alone, but under the action of $Ca(OH)_2$ or $CaSO_4$, its potential hydraulicity can be activated giving a slow hydraulic reaction. Activation by alkali metal compounds such as silicates or hydroxides gives rise to an intensive hydraulic reaction forming a strong, hardened body, the so-called alkali–slag cements. This type of material can serve not only as an additive for concrete, but as the main raw material for alkali–slag cementitious material.[33]
2. Additives with pozzolanic reactivity only, such as fly ash, metakaolinite, silica fume, diatomaceous earth, rice hush ash, zeolites, calcined gangue, lithium slag, tuff, pumice, and natural pozzolans. In these materials, the mass proportion of CaO is small, but there is a large amount of active SiO_2 and Al_2O_3. They have no hydraulicity or potential hydraulicity when used alone. Their activity is only realized at normal temperatures as a secondary reaction with Ca $(OH)_2$, generated during the hydration of cement (the pozzolanic reaction) to form calcium silicate and calcium aluminate hydrates with cementitious properties.
3. Additives with both potential hydraulicity and pozzolanic properties, such as high-calcium fly ash (where the mass proportion of CaO is 15% to 20%), liquidized slag, and sulfur consolidated slag. These additives not only have a significant amount of active SiO_2 and Al_2O_3,

but they are rich in CaO, although their content is far less than that in slags, but much more than in the second type of additives. Therefore, they have both potential hydraulicity and pozzolanic properties.

The classification given above is not absolute, as, for example, the first type of additives may also have certain pozzolanic reactivity.

The factors that influence the quality of the mineral admixtures are as follows.

4.1.1 Degree of amorphism

Minerals with the same chemical composition will have greater reactivity with $Ca(OH)_2$ under normal temperatures if they are more amorphous and less crystalline. Lian Hui-zhen[10] has suggested an index of activity rate calculated as follows:

$$K_a = \frac{\mathrm{m\ (active\ SiO_2) + m\ (active\ Al_2O_3)}}{\mathrm{m\ (total\ SiO_2) + m\ (total\ Al_2O_3)}}$$

and he obtained values for silica fume of 47.65%, for shale ash 32.43%, for zeolite 30.48%, for diatomaceous earth 19.02%, and for fly ash class III 12.29%.

4.1.2 Chemical composition of active mineral admixtures

Two materials, both with an amorphous structure but different chemical composition, can show differing activity; the content of CaO, SiO_2, and Al_2O_3 is the most significant factor influencing the activity. For the first and third types of additives, the most important factor is CaO content. The higher the CaO, the greater the potential hydraulicity of the additive and the more it can replace cement in concrete. Therefore, the activity rate is suitable only for the second type of additives to estimate activity.

For the first type of active mineral admixtures, the following factors are usually used for evaluation:

the basicity factor:

$$M = \frac{\mathrm{m\ (CaO) + m\ (MgO)}}{\mathrm{m\ (SiO_2) + m\ (Al_2O_3)}}$$

the mass factor:

$$K = \frac{\mathrm{m\ (CaO) + m\ (MgO) + m\ (Al_2O_3)}}{\mathrm{m\ (SiO_2) + m\ (MnO) + m\ (TiO_2)}}$$

When M > 1, the slag is called basic, when M = 1 it is neutral, while when M < 1 it is acidic. The higher the basicity and mass factors, the better the quality of the additive will be. The basicity factor for slag is usually between 0.78 and 1.35 and the mass factor is usually 1.43 to 2.45.

4.1.3 Fineness of active mineral admixtures

The finer the additive, the smaller its average particle size and the higher its specific surface, so more surface can take part in the reaction thus giving more strength. The specific surface area of silica fume is 15 to 25 m^2/g while that of ground fly ash and ground slag only reaches 0.7 to 0.8 m^2/g.

4.1.4 Water demand of active mineral admixtures

The water demand of active additives in concrete will have an effect on the fluidity of the mix. To keep the fluidity constant for an additive with higher water demand, the quantity of mixing water should be increased, which in turn increases the water/binder ratio, reducing the strength of the concrete. The water demand of an active mineral admixture is related to the compactness of the particles, to the roughness of the surface and impurity content, and, in particular, to the fineness of the material. An additive with high fineness will have a high water demand, as will those with a loose structure and a rough surface with large inner surface area, while those with a compact structure with a smooth surface will have low water demand. An additive originating from fuel ash with a high proportion of unburnt carbon will have high water demand and the durability of the concrete will be lower.

By using superplasticizers, the importance of water demand can be reduced. For example, silica fume and other superfine mineral admixtures will, with the help of a superplasticizer, serve as excellent active mineral admixtures and give an increase in fluidity (see Section 3.4).

4.2 A NEW METHOD OF ANALYSIS FOR THE POZZOLANIC EFFECT FOR SUPER-HIGH-STRENGTH HIGH PERFORMANCE CONCRETE AND AN EVALUATION OF THE ADDITIVE ACTIVITY

There are many methods for evaluating the activity of active additives. Apart from the activity rate method, the basicity factor method, and the mass factor method described above, there is the lime absorption value method, the pozzolan activity diagram method, and the commonly used compressive strength ratio method,[34] which compares the 28-day compressive strength of a cement mortar with 30% of an active mineral admixture with a pozzolanic character to the 28-day compressive strength of a pure Portland cement mortar. When the ratio is more than 0.62, such an additive can be considered as pozzolanic. This is because the compressive strength ratio of a non-active additive, ground sand, is 0.62. The greater this value, the greater the activity of the pozzolanic material.

In 2000, China published a National Standard (GB/T 18064-2000) *Granulated Blast Furnace Slag used for Cement Concrete*, in which the activity index was used to evaluate the activity of slag. The standard base mortar mix used 450 g of 525 Portland cement, 1350 g standard Chinese ISO sand with a water/cement ratio of 0.5. To test the slag, 225 g of 525 Portland cement with 225g (50%) of slag powder (specific surface area >3500 cm^2/g), 1350 g of ISO standard sand, was used to prepare mortar samples with a water/binder ratio of 0.5. The percentage of the ratio of 7-day strength and 28-day strength for these two specimens was taken as the activity index. Although this standard is based on an international one, the essence of evaluating the activity of the slag in this way is based only on the concept of compressive strength ratio, and it is suitable only for slag.

As pointed out by Academician Wu Zhongwei and Professor Lian Huizhen,[10] "it was hoped to find a rapid and reliable method for evaluating the activity of the mineral admixture, but we failed." For example, the lime absorption value method did not reflect the quality of cement and concrete after addition of active additives; the pozzolan activity diagram method only reflected whether the additive had activity, but could not judge correctly the contribution the additive made to the strength of cement and concrete and *only* described the pozzolanic effect behavior qualitatively. Thus, the author suggests that a quantitative analysis method be based on the concept of specific strength.

Any factor which influences the activity of a mineral admixture will be expressed in its pozzolanic reactivity with cement and finally in the strength index of concrete; therefore, this method is still based on the experimental data of mechanical resistance.

In a concrete-containing active mineral admixture, the binder is composed of both the cement and the additive. While the cement has independent hydraulicity, the active mineral admixture does not, so its contribution to the strength of concrete is based on the secondary pozzolanic reaction, by the $Ca(OH)_2$ and other hydrates produced during the hydration of cement or by stimulating the potential hydraulicity and the water-reducing effect and filling effect of the additive. The sum of all these effects is called the pozzolanic effect. Thus, the strength of concrete containing an additive is composed of two parts, namely, the strength contribution from cement hydration and the strength contributed by the pozzolanic effect of the additive. Because variations in the additive replacement will significantly affect the absolute value of concrete strength, it cannot be used as a basis for comparison. Thus, the author suggests the contribution of a cement content unit (1% of cement content) to the strength of concrete (in the case of a cement with an additive, the clinker content unit is used). This is called by the author the *specific strength of cement content in concrete*, or, in short, the *specific strength of concrete* (it is different from the specific strength of material mass given in Table 1.2). The specific strength of concrete with an additive (R_{sa}, MPa) is as follows:

$$R_{sa} = \frac{R_a}{q_0} \tag{4.1a}$$

where

R_a is the absolute value of strength of concrete with additive, in MPa
q_0 is the mass percentage of cement in the concrete containing additive, %

In the basic concrete without additive for comparison, its mass percentage of cement is 100% and its specific strength (R_{sc}, MPa) given by:

$$R_{sc} = \frac{R_c}{100} \tag{4.1b}$$

where R_c is the absolute value of the basic concrete strength in MPa.

Though the cement content is reduced in the concrete containing an additive due to the pozzolanic effect, R_a is often more than R_c, and R_{sa} more than R_{sc} so the difference between the two values is the specific strength contributed by pozzolanic effect, called by the author *the specific strength of pozzolanic effect* (R_{sp}, MPa), where

$$R_{sp} = R_{sa} - R_{sc} \tag{4.2}$$

Consequently, we can define the specific strength factor K as:

$$K = \frac{R_{sa}}{R_{sc}} \tag{4.3}$$

At the same time, the author has suggested a strength contribution rate of pozzolanic effect (P_a%), which expresses quantitatively the magnitude of the contribution of the pozzolanic effect of the additive to the strength of concrete:

$$P_a = \frac{R_{sp}}{R_{sa}} \times 100\% \tag{4.4}$$

Similarly, we can define the strength contribution rate of hydration (P_h,%):

$$P_h = \frac{R_{sc}}{R_{sa}} \times 100\% \tag{4.5}$$

It is obvious $P_a + P_h = 100\%$.

If the strength contribution rate of the pozzolanic effect P_a is divided by the mass percentage of the active mineral admixture, q, we will obtain the rate of strength contribution of the pozzolanic effect offered by a unit of active mineral admixture (1% of active mineral admixture). The author defines this index as the activity index, A, of this particular additive (which is different from the activity index given in GB/T 18046-2000):

$$A = \frac{P_a}{q} \tag{4.6}$$

If the strength contribution to concrete from 1% of additive is compared with that of 1% of cement, when A > 1, the contribution from the additive is greater than that from cement; when A = 1, the contribution from the additive is the same as that from cement; and when A < 1, the contribution of the additive is less.

When two or more active mineral admixtures are used in combination, the combined strength contribution rate of pozzolanic effect P_{com} can be calculated. In general, for combined admixtures, no more than three types of additive are used, therefore:

$$P_{com} = P_t - P_1 - P_2 - P_3 = P_t - q_1A_1 - q_2A_2 - q_3A_3 \tag{4.7}$$

where

P_t is the total contribution rate of pozzolanic effect to the strength of combined additive,%.

P_1, P_2, P_3 are the strength contribution rates of pozzolanic effect for each additive, respectively.

A_1, A_2, A_3 are activity indexes of additives 1, 2, 3, respectively.

q_1, q_2, q_3 are mass percentages of additives 1, 2, 3, respectively.

The function and advantage of this method are as follows:

1. The result is quantitative so these data can be used for quantitative analysis of the pozzolanic effect. This is exact and reliable, and is the greatest advantage of this method.
2. The method is widely adaptable and suitable for identification and analysis of the activity of all mineral admixtures, allowing the activity

of additives from different sources and types to be compared through the activity index,[30] which cannot be obtained by any other method.
3. Using this method, behavior of the pozzolanic effect of various mineral admixtures can be analyzed for different replacements and different ages, which also cannot be done by any other method.
4. The contribution rate of the various components in the binder to the strength of concrete can be analyzed as well as the strength composition of the concrete, leading to a deeper understanding of the origin of concrete strength, and offering a tool for analyzing the composition of the concrete strength.
5. This method is simple and does not require other chemical, physical, and microscopic tests and it needs only one more specimen of basic concrete to be prepared under the same conditions.

4.3 ANALYSIS OF THE POZZOLANIC EFFECT OF ACTIVE MINERAL ADMIXTURES

As described in the previous section, the activity of a mineral admixture can be expressed as its ability to react as a pozzolan with the cement and finally in the strength index of the concrete. Here, an analysis will be carried out using the new method described above and based on experimental data.

In Table 4.1, the 28-day compressive strengths of concretes containing a slag additive at replacement levels of 20% to 60% obtained by Zhang Xiong et al.[24] are given, together with analysis of the results. It can be seen from the table that

1. With increasing slag replacement of cement, the compressive strength initially increased up to a replacement level of 40% when the maximum strength was reached. Increasing the replacement level up to 50% to 60% gave rise to a strength decrease. From the point of view of absolute strength, there is an optimum slag replacement when strength reaches the maximum.
2. With increasing slag addition, the strength contribution due to the pozzolanic effect increased continuously, from an addition of 20%, where the contribution rate was 32.1%, to 60%, where the contribution rate increased to 62.6%. It is clear how large and important the pozzolanic effect is to the overall strength. The large pozzolanic effect of the slag is due to its chemical composition, with the mass percentage of CaO in slag reaching 35% to 48%, second only to clinker (63% to 65%). It has potential hydraulicity, so it can be added in large amounts (up to 70% to 80%, although the absolute strength value of concrete is reduced). Therefore, the strength contribution rate of the pozzolanic effect will be very high.

Table 4.1 Strength contribution rate of the pozzolanic effect and the activity index of high performance concrete with a slag additive

Mix proportion		*1*	*2*	*3*	*4*	*5*	*6*
Mass percentage of binder in concrete (%)	Cement	100	80	70	60	50	40
	Slag	0	20	30	40	50	60
28-day compressive strength of concrete, R_c, R_a (MPa)		73	86	92	96	91	78
Specific strength of basic concrete, R_{sc} (MPa)		0.730					
Specific strength of concrete with additive, R_{sa} (MPa)			1.075	1.314	1.600	1.820	1.950
Specific strength of pozzolanic effect, R_{sp} (MPa)			0.345	0.584	0.870	1.090	1.220
Specific strength factor, K			1.473	1.800	2.192	2.493	2.671
Strength contribution rate of pozzolanic effect, P_a (%)		0	32.1	44.4	54.4	59.9	62.6
Strength contribution rate of hydration, P_h (%)		100	67.9	55.6	45.6	40.1	37.4
Activity index, A			1.61	1.47	1.36	1.20	1.04

3. The activity index, A, of the slag additive reduced with the addition rate, falling from 1.61 at 20% replacement down to 1.04 at 60% replacement. However, the index, A, still remained greater than 1, which means that at a slag replacement of 60%, the strength contribution of 1% of slag was still more than that of 1% of cement.

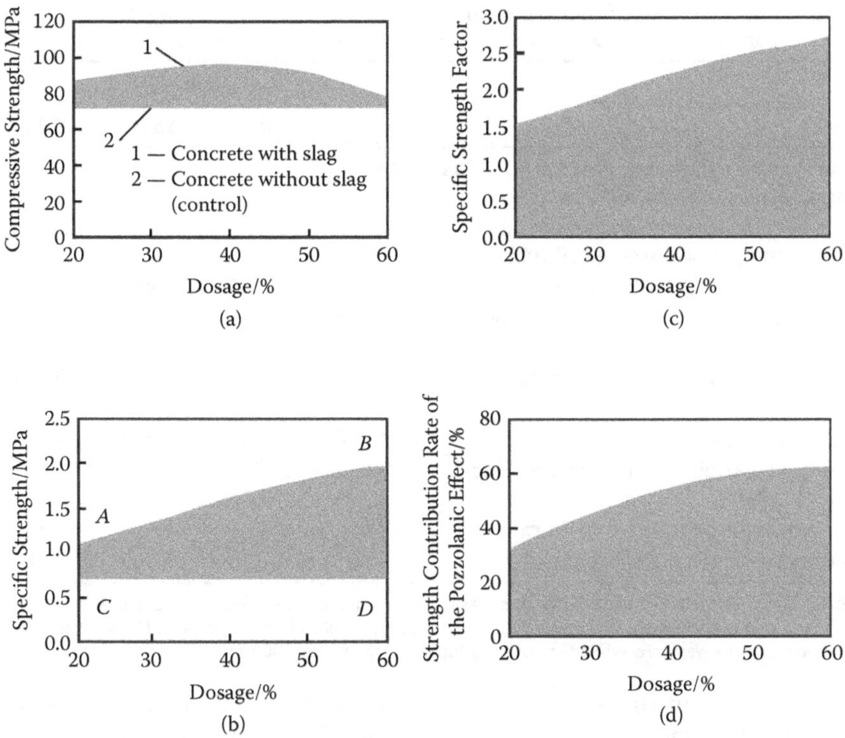

Figure 4.1 The relationship between the strength, specific strength, and specific strength factor as well as the strength contribution rate of the pozzolanic effect of concrete with the slag additive in respect to the additive addition.

Figure 4.1 is constructed from the data in Table 4.1 and illustrates the relationship between the strength, specific strength, and specific strength factor as well as the strength contribution rate of the pozzolanic effect of the slag additive in concrete with respect to the additive addition. It can be seen from Figure 4.1(b) that for the addition range discussed, an increase of addition gives a continuous rise in the specific strength of the concrete. The area between the curve AB and the horizontal line for the base concrete CD shows the region of pozzolanic effect for the range of addition, discussed as a diagram of pozzolanic effect for the specific strength of concrete with the replacement of additive.

Table 4.2 illustrates the specific strength determined from data obtained from tests on super-high-strength silica fume concretes by the author and his colleagues. The specific strength of the pozzolanic effect, the specific strength factor, the strength contribution rate of pozzolanic effect, and the strength contribution rate of hydration and activity index of silica fume

Table 4.2 Strength, specific strength, strength contribution rate of pozzolanic effect, and activity index of super-high-strength silica fume concrete

Index	Age *3 d*	*28 d*	*56 d*	*90 d*
Basic concrete (Cement: silica fume = 100:0) compressive strength, R_c/MPa	64.1	76.3	81.4	93.8
Silica fume concrete (Cement: silica fume = 90:10) compressive strength, R_a/MPa	68.7	106.2	112.0	126.0
Specific strength of basic concrete, R_{sc}/MPa	0.641	0.763	0.814	0.938
Specific strength of silica fume concrete, R_{sa}/MPa	0.763	1.169	1.244	1.400
Specific strength of pozzolanic effect, R_{sp}/MPa	0.122	0.406	0.403	0.464
Specific strength factor, K	1.190	1.532	1.528	1.496
Strength contribution rate of pozzolanic effect, P_a/%	16.0	34.7	34.6	33.1
Strength contribution rate of hydration, P_h/%	84.0	65.3	65.4	66.9
Activity index of silica fume, A	1.60	3.47	3.46	3.31

Note: The water binder ratio of concrete was 0.25, the dosage of S-20 water-reducing agent was the same, the slump of basic concrete was 110 mm, the slump of silica fume concrete 235 mm, the spread of basic concrete was zero while that of silica fume concrete was 580 mm.

are also given in this table. Figure 4.2, which is based on the data given in Table 4.2, shows the variation of strength, specific strength, specific strength of the pozzolanic effect, and the strength contribution rate of pozzolanic effect with the age of concrete. From Figure 4.2(c), the variation of specific strength of the pozzolanic effect of the concrete with its age is shown, in what may be called a *diagram of the pozzolanic effect of concrete with age*. The activity index of the additive also varied with the age of concrete. The silica fume had high activity after 3 days of hardening when the activity index reached 1.60. From 3 to 28 days, the pozzolanic effect developed quickly, so by the age of 28 days, the pozzolanic effect was the strongest with the strength contribution rate of pozzolanic effect reaching its maximum, 34.7% (activity index 3.47), and then reducing slowly.

4.4 ANALYSIS OF THE STRENGTH COMPOSITION OF SUPER-HIGH-STRENGTH HIGH PERFORMANCE CONCRETE

The author and his collaborators have prepared samples of super-high-strength concrete using Kunming 625 Portland cement, superplasticizer,

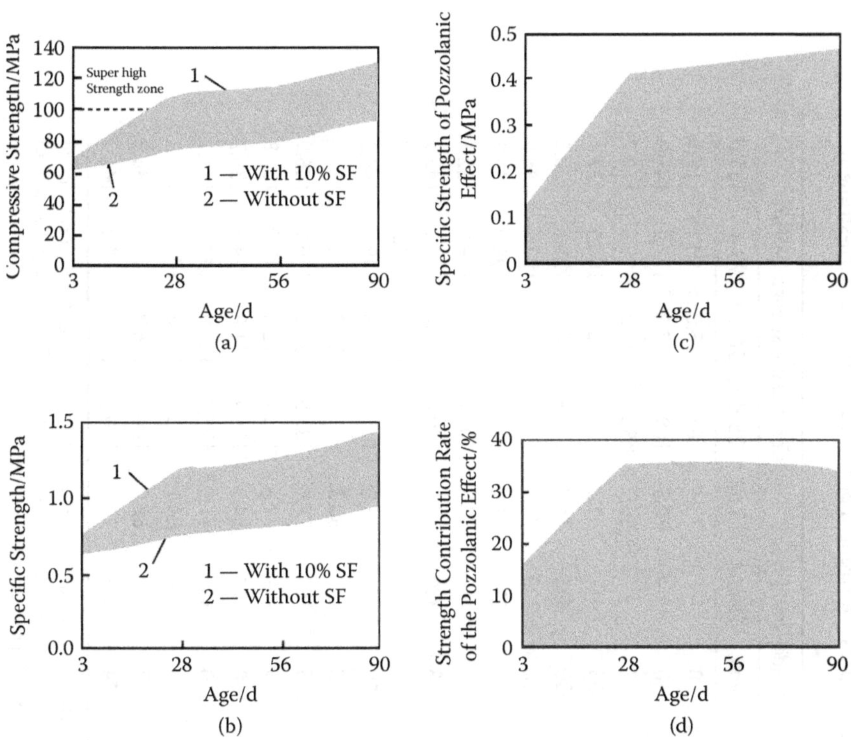

Figure 4.2 The strength, specific strength, specific strength of the pozzolanic effect, and strength contribution rate of pozzolanic effect for super-high-strength silica fume concrete.

and an active mineral admixture.[35,36] The active mineral admixtures used, apart from silica fume, superfine slag, and superfine fly ash, were new types of active mineral admixtures, X powder and Y powder. Besides single additions, combined additions of two or three types of additives were used. In order to avoid any influence due to level of dosage, the total replacement of additive in all mixes was the same (20%). The mix proportions, along with fluidity after mixing and strengths at different ages, are given in Table 4.3. The specific strength, specific strength of pozzolanic effect, strength contribution rate of pozzolanic effect, strength contribution rate of hydration, activity index of mineral admixture, and strength contribution rate of combined effect, obtained by calculation from the 28-day strength data in Table 4.3, are given in Tables 4.4 and 4.5.

Table 4.3 Composition of binder, fluidity of mix, and compressive strength of concrete for super-high-strength high performance concrete

	Composition of binder							*Fluidity (mm)*		*Compressive strength (MPa)*				
No.	*C*	*SF*	*SL*	*FA*	*X*	*Y*	*W/B*	*Slump*	*Spread*	*3 d*	*28 d*	*56 d*	*90 d*	*365 d*
1	100						0.22	0	—	73.4	87.3	95.0	98.1	117.1
2	80	20					0.22	140	—	78.3	109.4	117.8	127.2	146.9
3	80		20				0.22	8	—	56.2	95.0	97.2	99.6	118.7
4	80			20			0.22	105	—	52.3	91.7	96.4	100.5	127.1
5	80	10	10				0.22	230	445	74.8	111.4	118.5	125.6	138.6
6	80	10		10			0.22	208	393	73.8	114.9	130.3	132.0	145.9
7	80	5	7.5	7.5			0.22	222	391	66.2	117.4	122.1	129.3	132.2
8	80				20		0.22	167	—	50.2	105.2	119.9	124.8	132.5
9	80	10			10		0.22	218	380	77.4	120.0	124.7	133.1	143.9
10	80					20	0.22	20	—	70.2	98.9	110.0	114.5	132.3
11	80	10				10	0.22	159	—	75.0	110.1	126.6	129.9	144.3
12	100						0.20	0	—	80.3	95.7	104.0	113.3	127.3
13	80	20					0.20	215	354	77.4	132.5	134.8	140.5	146.9
14	80		20				0.20	9	—	72.5	106.7	112.4	118.9	137.8
15	80			20			0.20	198	300	54.9	106.2	111.8	124.5	138.7
16	80	10		10			0.20	210	330	77.5	127.4	130.8	142.3	151.1
17	80	10	10				0.20	163	284	84.1	125.0	135.9	141.1	151.3
18	80			10	10		0.20	7	—	68.9	106.2	116.9	119.6	135.9
19	80	5	7.5	7.5			0.20	174	301	73.9	120.6	124.8	130.3	146.3

Notes: (1) The binder content is 600 kg/m^3 in mix No. 1 through No. 11, and 650 kg/m^3 in mix No. 12 through No. 19. (2) Crushed limestone, medium sand (see Section 6.2), and naphthalene sulphonate superplasticizer S-20, added in fixed mass percentage of binder. (3) C = cement; SF = silica fume; SL = slag; FA = fly ash; X = X powder; Y = Y powder.

Table 4.4 Pozzolanic effect and composition of strength of super-high-strength concrete at 28 days (cementitious material 600 kg/m^3)

		Proportion of cementitious material (%)							*Index of pozzolanic effect*						*Composition of strength/MPa*							
								28-day strength (MPa)							*From hydration*	*Contribution from pozzolanic effect*						*Complex effect*
		C	*FA*	*SF*	*SL*	*X*	*Y*		R_s *(MPa)*	R_{sp} *(MPa)*	P_a *(%)*	P_h *(%)*	A_i	P_{com} *(%)*		*Total*	*FA*	*SF*	*SL*	*X*	*Y*	
		Symbols																				
Type of concrete	*No.*	q_o	q_1	q_2	q_3	q_4	q_5	R	R_s	R_{sp}	P_a	P_h	A_i	P_{com}	RP_h	RP_a	RA_1q_1	RA_2q_2	RA_3q_3	RA_4q_4	RA_5q_5	RP_{com}
Basic	1	100						87.3	0.873	0	0	100	—		87.3							
Concrete with additive	2	80	20					91.7	1.140	0.273	23.82	76.18	1.19 (A_1)		69.8	21.9	21.9					
	3	80		20				109.4	1.368	0.495	36.18	63.82	1.81 (A_2)		69.8	39.6		39.6				
	4	80			20			95.0	1.188	0.315	26.56	73.48	1.33 (A_3)		69.8	25.2			25.2			
	5	80	10	10				114.9	1.436	0.563	39.21	60.79		9.21	69.8	45.1	13.7	20.8				10.6
	6	80	—	10	10			114.4	1.393	0.520	37.33	62.67		5.93	69.8	41.6		20.0	14.8			6.6
	7	80	7.5	5	7.5			117.4	1.468	0.595	40.53	59.47		12.51	69.8	47.6	10.5	10.6	11.7			14.8
	8	80				20		105.2	1.315	0.442	33.61	66.39	1.68 (A_4)		69.8	35.4				35.4		
	9	80		10		10		120.2	1.500	0.627	41.80	58.20		6.90	69.8	50.2		21.7		20.2		8.3
	10	80					20	98.9	1.235	0.363	29.37	70.63	1.47 (A_5)		69.8	29.0					29.0	
	11	80		10			10	110.1	1.375	0.502	36.51	63.49		3.51	69.8	40.2		19.9			16.4	3.9

Source: Pu Xincheng, Analysis on Strength Composition of Super High Strength High Performance Concrete, *Concrete and Cement Products*, 1999 (1) (in Chinese).

Note: $R_s = R/q_i$; $R_{sp} = R_{sa} - P_{sc}$; $P_a = (R_{sp}/R_{sa}) \times 100\%$; $A_i = P_{ai}/q_i$; $P_{com} = P_t - \Sigma A_i q_i$; C = cement; FA = fly ash; SF = silica fume; SL = slag; X = X powder; Y = Y powder; R_s = specific strength of concrete; R_{sa} = specific strength of concrete with additive; R_{sc} = specific strength of basic concrete; R_{sp} = specific strength of pozzolanic effect of concrete; P_a = strength contribution rate of pozzolanic effect; P_h = strength contribution rate of hydration; A_i = activity index of single additive; P_{ai} = strength contribution rate of pozzolanic effect of single additive; P_t = total strength contribution rate of additives in combination; q_i = replacement of single additive; P_{com} = strength contribution rate of complex effect.

Table 4.5 Pozzolanic effect and strength composition of super-high-strength concrete at 28 days (cementitious material 650 kg/m³)

Type of concrete	No.	Proportion of cementitious material (%) C	FA	SF	SL	28-day strength (MPa)	Index of pozzolanic effect R_s (MPa)	R_{sp} (MPa)	P_a (%)	P_h (%)	A_i	P_{com} (%)	Composition of strength/MPa: From hydration	Contribution from pozzolanic effect: Total	FA	SF	SL	Complex effect
		q_0	q_1	q_2	q_3	R	R_s	R_{sp}	P_a	P_h	A_i	P_{com}	RP_h	RP_a	RA_1q_1	RA_2q_2	RA_3q_3	RP_{com}
Basic	12	100	—	—	—	95.7	0.957	0	0	100	—	—	95.7					
Concrete with additive	13	80	20	—	—	106.2	1.328	0.371	27.94	72.06	1.40 (A_1)	—	76.5	29.7	29.7	—	—	—
	14	80	—	20	—	132.5	1.656	0.699	41.21	57.79	2.11 (A_2)	—	76.5	55.9	—	55.9	—	—
	15	80	—	—	20	106.7	1.334	0.377	28.26	71.74	1.41 (A_3)	—	76.5	30.2	—	—	30.2	—
	16	80	10	10	—	127.4	1.593	0.636	39.92	60.08	—	4.82	76.5	50.9	17.8	26.9	—	6.2
	17	80	—	10	10	125.0	1.563	0.606	38.77	61.23	—	3.57	76.5	48.5	—	26.4	17.6	4.5
	18	80	10	—	10	106.2	1.328	0.371	27.94	72.06	—	–0.16	76.5	29.7	14.9	—	15.0	–0.2
	19	80	7.5	5	7.5	120.6	1.508	0.511	36.54	63.46	—	4.91	76.5	44.1	12.7	12.7	12.8	5.9

Source: Pu Xincheng, Analysis on Strength Composition of Super High Strength High Performance Concrete, *Concrete and Cement Products*, 1999 (1) (in Chinese).

Note: $R_s = R/q_i$; $R_{sp} = R_{sa} - P_{sc}$; $P_a = (R_{sp}/R_{sa}) \times 100\%$; $A_i = P_{ai}/q_i$; $P_{com} = P_t - \Sigma A_i q_i$; C = cement; FA = fly ash; SF = silica fume; SL = slag; X = X powder; Y = Y powder; R_s = specific strength of concrete; R_{sa} = specific strength of concrete with additive; R_{sc} = specific strength of basic concrete; R_{sp} = specific strength of pozzolanic effect of concrete; P_a = strength contribution rate of pozzolanic effect; P_h = strength contribution rate of hydration; A_i = activity index of single additive; P_{ai} = strength contribution rate of pozzolanic effect of single additive; P_t = total strength contribution rate of additives in combination; q_i = replacement of single additive; P_{com} = strength contribution rate of complex effect.

It can be seen from Table 4.3 through Table 4.5 that

1. The additive used is important in preparing super-high-strength concrete. The basic concrete without an additive only reached a 28-day strength of 87.3 MPa with the concrete having a slump of zero. With the addition of an additive, the 28-day compressive strength was enhanced and the fluidity increased by different amounts, especially when silica fume and X powder were combined giving rise to a compressive strength of 120 MPa with a slump of 218 mm.
2. The strength contribution rate of the pozzolanic effect is high, with a minimum of 23.82% (mix No. 4) and a maximum of 41.80% (mix No. 9).
3. When the activity indexes of the five additives are compared, the sequence is as follows: silica fume (1.81) > X powder (1.68) > Y powder (1.49) > slag (1.33) > fly ash (1.19). Consequently, there is a bright future for this new type of mineral admixture X powder and Y powder, which can give a super-high-strength high performance concrete.
4. The combination of additives can have a complex effect. In this study, the contribution of the combined effect to strength is in range of 3.51% to 12.51%, with the best result coming from a combination of three additives—silica fume, slag, and fly ash—which reached 12.51%. This complex effect cannot be neglected. It should be pointed out though that any combination always requires silica fume to ensure the effect occurs. For example, combining slag with fly ash (mix No. 18) did not produce the complex effect because their mean particle sizes are of the same order.

This complex effect is mainly based on the physical effect of compact filling of the binder. The average particle size of cement is of the order of 10 μm, that of superfine ground additive around 1 μm, while for silica fume it is of the order of 0.1 μm. Because of this large difference in particle size, the particles can be compactly filled with little repulsion. During concrete placement, the mold can be readily filled with a flowable mix leading to increased compaction of hardened cement paste.[30] The pozzolanic activity of fly ash and slag is well known to be much less than that of silica fume. Thus, in mix No. 7, with only 5% silica but with 15% substitution of fly ash and slag, the strength contribution rate of pozzolanic effect reached 40.53%, and the absolute value of the strength was higher than that of concrete with only silica fume added. This cannot be explained from the viewpoint of chemical reactivity. As silica fume is expensive and its supply limited, so the approach of a combination of additives has great promise for the future.

When the strength contribution rates of cement hydration and the pozzolanic effect of each mix are known, then in the case of combined additives, where the strength contribution rates of the additive and the complex effect are known, the strength values contributed by hydration; the total pozzolanic effect for both the single additive and the complex effect of combined additives can be calculated by multiplying the strength value of the concrete with those strength contribution rates given above. These results are given in Tables 4.4 and 4.5. From these calculations and analyses, we can clearly know the strength contribution rate of materials and their contribution to concrete strength, allowing us to analyze the composition of the concrete strength. This new method for analyzing the pozzolanic effect presented by the author can be used not only for determining the activity of different mineral admixtures, but it can also serve as a tool for the study and analysis of the strength composition of concrete.

It should be emphasized though that only correct experimental data can guarantee the accuracy of the results of analysis. To ensure that these analyses are correct, apart from the type of additive, the raw materials of concrete, the mix proportion parameters (binder content, total amount of additive replacement, water/binder ratio, sand percentage, dosage of water-reducing agents, maximum size of coarse aggregate, etc.), together with curing conditions, age, and so forth, must be kept constant for all mixes. Otherwise, the data cannot be compared.

Chapter 5

Raw materials

5.1 PORTLAND CEMENT

Portland cement is manufactured as follows. A calcareous material (limestone) and a siliceous material (clay, shale) are ground together, usually with the addition of an iron powder in certain proportions, to form a raw meal (for dry technology), or a slurry (wet technology), which is fed into a rotary kiln and calcined at temperatures up to 1450°C. Partial melting occurs to form a clinker which is rapidly cooled. This can be stored before it is mixed with a certain amount of gypsum and interground. The final product is called Portland cement.

There are four key mineral components in Portland cement: tricalcium silicate ($3CaO{\cdot}SiO_2$, C_3S), β-dicalcium silicate ($2CaO{\cdot}SiO_2$, C_2S), tricalcium aluminate ($3CaO{\cdot}Al_2O_3$, C_3A), and tetracalcium aluminoferrite ($4CaO{\cdot}Al_2O_3{\cdot}Fe_2O_3$, C_4AF). Table 5.1 illustrates quantitatively the basic properties of these four minerals. By carefully selecting the proportions of raw feed, the mass percentage of these four minerals in cement can be varied giving rise to cements with different properties. For example, by raising the mass percentage of C_3S, a high strength cement can be produced; enhancing the mass percentage of C_3S and C_3A gives a rapid hardening and high early-strength cement; lowering the mass percentage of C_3S and C_3A produces a low heat Portland cement. Figure 5.1 shows the curves for strength development of these four minerals. The results for the heat of hydration are given in Table 5.2.

The requirements for cement that is used as the basic binding material of super-high-strength high performance concrete preparation are as follows:

1. The cement grade should be high strength.
2. As the proportion of cement in super-high-strength concrete is rather high, during hardening of the concrete, the hydration heat will also be high resulting in temperature of the concrete increasing and this may lead to large differences in temperature between the inside and outside of the concrete leading to temperature cracking. Therefore, in

Table 5.1 Characteristics of minerals in Portland cement clinker[2]

Minerals		*C_3S*	*C_2S*	*C_3A*	*C_4AF*
Character	Hydration rate	Fast	Slow	Fastest	Medium
	Heat of hydration	High	Little	Highest	Low
	Strength	High	Early low, later high	Low	Low
Mass percentage (%)		37–60	15–37	3–12	8–15

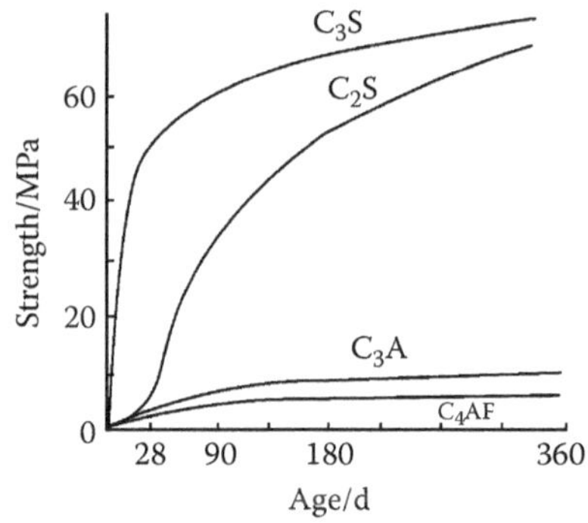

Figure 5.1 Strength development of clinker minerals of Portland cement.

Table 5.2 Heat of hydration for Portland cement in J/g

	Age/d					
Mineral	*3*	*7*	*28*	*90*	*180*	*Complete hydration*
C3S	406	460	485	519	565	669
C2S	63	105	167	184	209	331
C3A	590	661	874	929	1025	1063
C4AF	92	251	377	414	—	569

Source: Zhou Shiqiong, *Building Materials*, Beijing: China Railway Press, 1999 (in Chinese).

northern European countries, a special cement for high strength high performance concretes has been developed. It is characterized by a high mass percentage of C_2S with

$$\frac{m(C_2S)}{m(C_3S)+m(C_2S)} \geq 0.5$$

In China, a high belite cement has been developed, where the mass percentage of βC_2S reaches 50.67. The grade of cement mortar reaches a

28-day compressive strength up to 63 MPa (grade 525 [GB 1344-92]).[38] Such cement can be used as the binder for super-high-strength concrete.

3. The mass percentages of both alkali and C_3A, plus C_4AF should be low as well to ensure good compatibility between the cement and the superplasticizer. At the same time, anhydrite should not be used as a retarder for the cement.
4. Low water is required for the cement to ensure sufficient fluidity when lowering the water/cement ratio. Certainly, it is difficult to lower the water requirement for ordinary Portland cement. In the 1980s in the former Soviet Union, a low water requirement binding material was produced by intergrinding Portland cement and superplasticizer, allowing the water requirement to be lowered from 24.7% down to 17%. The idea was that before the cement hydrated, the superplasticizer would be adsorbed onto the surface of the cement particles, avoiding the adsorption of a large part of superplasticizer by the aggregate, thus reducing the plasticizing effect. In 1994, the Russian Research Institute of Cement developed a new type of low water requirement cement. It was characterized by raising the mass percentage of C_3S in the clinker to 70% using a special catalyst to lower the energy consumption of calcination and grinding. The cement was obtained by intergrinding clinker, gypsum, superplasticizer, and a fine additive to give a cement that would produce a 170-MPa super-high-strength concrete.[107]

5.2 ACTIVE MINERAL ADMIXTURE

As described above, the active mineral admixture is a significant element in preparing super-high-strength high performance concrete. From the experience of the author,[31] even using a large amount of high grade cement and superplasticizer but without an active mineral admixture, it is not possible to produce a super-high-strength high performance concrete with a 28-day strength more than 100 MPa, and zero slump. Not all active mineral admixtures can be used to prepare super-high-strength high performance concrete. Only the best, such as silica fume, superfine fly ash, superfine slag, superfine X powder, superfine Y powder, superfine calcined kaolin, and superfine rice husk ash, are suitable. In normal construction practice, to prepare concrete with over 80-MPa compressive strength, silica fume is needed.

5.2.1 Silica fume

Silica fume is a superfine powder collected from the chimneys of ferro-silicon or metallic silicon smelters. Powdered quartz is reduced to Si by carbothermal reduction with coke in an electric kiln. SiO_2 is formed by steam oxidation at low temperatures and coagulates into amorphous ball-like particles. The BET surface area measured by nitrogen adsorption reaches 150,000 to 250,000 m^2/kg and the mean particle size is 0.1 to 0.2 μm. The mass percentage of SiO_2 in silica fume depends on the type of alloy being produced and varies from 90% to 98% (such as from metallic silicon, ~75% from FeSi and as low as 25% to 54% from SiMn, or CaSi products. Users should know the source and quality of the fume since silica fume used for preparing super-high-strength concrete should have a SiO_2 mass percentage of more than 90%.

5.2.2 Blast furnace slag

During the manufacture of iron in a blast furnace, in addition to iron ore and fuel (coke) in the furnace, limestone or dolomite are added as a flux to lower the temperature of melting in the furnace and to scavenge the impurities. The calcium and magnesium oxides formed at high temperature combine with the nonferrous material (waste stone) from iron ore and ash from coke to form mainly calcium and magnesium silicates and aluminosilicates. The fluidized slag with its lower density of 1.5 to 2.2 floats on the surface of molten iron. The slag is tapped regularly from the blast furnace and cooled rapidly (water quenched or with a pressurized air jet [air quenched]) to form glassy granules and is called *granulated blast furnace slag*, or *water quenched blast furnace slag* when granulated by water. Such granulated slag when finely ground serves as an active mineral admixture. If the same slag is discharged into a ladle for transport and allowed to cool slowly, it crystallizes and cannot be used as an additive.

The main chemical components in slag are CaO, SiO_2, and Al_2O_3, making up a total mass percentage of more than 90% with small amounts of MgO, Fe_2O_3, TiO_2, MnO, and so forth present. The activity of the slag depends mainly on the relationship between the key oxides: the higher the basicity factor, M, the larger the potential hydraulicity and the higher the reactivity.

$$M = \frac{\mathrm{m(CaO)} + \mathrm{m(MgO)}}{\mathrm{m(SiO_2)} + \mathrm{m(Al_2O_3)}}$$

Usually, M lies between 0.78 and 1.35. Basic slag with $M > 1$ is the most suitable as an additive for cement. However, an acidic slag with $M < 1$ is also suitable for preparing super-high-strength concrete acting as an active mineral admixture. This is because in an acidic slag, the mass percentage of SiO_2 is high allowing the pozzolanic reaction to occur during hardening of super-high-strength concrete. Slag as a mineral admixture for super-high-strength concrete should be ground to a very high specific surface area of more than 700 m^2/kg.

5.2.3 Fly ash

Fly ash is the dust collected from the stack discharge from the boiler during combustion of powdered coal (<100 μm) at a power plant. During the high temperature combustion, the clay minerals in the coal powder melt at the high temperature of the furnace (1100 to 1500°C). Spherical microparticles formed by the action of surface tension and rapidly cooled when discharged out of the boiler form 0.5- to 250-μm glassy silicate microspheres or cenospheres and can contain iron and aluminum, or porous glassy microparticles. In addition, small crystals of mullite, quartz, magnetite, hematite, and unburned carbon particles form. Apart from the fly ash, coarse bottom ash is also discharged from the bottom of the boiler, making up about 15% to 20% of the total ash.

There are two ways of discharging fly ash from the power plant: wet or dry discharge. In the case of wet discharge, the fly ash is transported by high-pressure water to a large distant ash storage site. In order to utilize wet ash, it needs to be dewatered and heated to dryness, so the method is seldom used. With the widespread utilization of fly ash, nowadays most fly ash is produced by the dry method and transported by a Fuller-Kinyon pump or a pneumatic pump into a fly ash silo and packed into bags or delivered pneumatically to the customer.

The collection method of fly ash can be divided into mechanical or electrical. The mechanical dust collection is less efficient, as there is a smaller percentage of fine particles. The electrostatic dust collector is often composed of two or more electric fields in series, one collecting coarse dust and others collecting successively finer ash.

The main chemical components of fly ash are SiO_2, Al_2O_3, and Fe_2O_3 (Table 5.3). According to the China National Standard, *Fly Ash for Cements and Concretes* (GB 1596-91), fly ash can be divided into three classes based on its chemical composition. Fly ash that can be used as an active additive for cement belongs only to the first and second classes. For the first class, the loss of ignition should be no more than 5%, and for the second class, no more than 8%. In both cases, the sulphur trioxide content should be no more than 3% with a 28-day compressive strength ratio of no

Table 5.3 Chemical components of fly ash

Oxides	SiO_2	Al_2O_3	Fe_2O_3	CaO	MgO	SO_3	Na_2O	K_2O	*Loss of ignition*
Range of mass percentage	38.86–59.68	16.49–35.38	1.46–15.43	0.77–10.33	0.70–1.85	0.06–1.05	0.24–1.10	0.74–2.35	1.23–23.5
Mean mass percentage	50.76	27.20	7.00	2.78	1.15	0.33	0.46	1.32	8.19

Source: Shen Dansheng and Wu Zhengyan, *Design of Modern Concrete*, Shanghai: Shanghai Scientific and Technical Literature Press, 1987 (in Chinese).

less than 75% and 65%, respectively. If the fly ash is to serve as an additive to mortar or concrete, the fineness of I, II, and III class ashes (measured by the residue on a 45-μm sieve with square holes) should be no more than 12%, 20%, and 45%, respectively, with water requirements of no more than 95%, 105%, and 115%; loss of ignition of no more than 5%, 8%, and 15%; and a sulphur content of no more than 3% for all classes.

Unrefined fly ash cannot be used as an additive for super-high-strength concrete but needs processing by grinding (I or II class) to a specific surface over 700 m^2/kg for use in super-high-strength concrete.

5.3 CHEMICAL ADMIXTURES

Chemical admixture is compound added into concrete at less than 5% of mass of binding material (cement + ground mineral admixture) to improve the properties of fresh and hardened concretes. In mix proportions of concrete, it does not substitute for the cement and so it is not included in the mass of cement. Examples are water-reducing agents and superplasticizers, retarders, shrinkage-reducing agents, durability-improving agents, air-entraining agents, accelerators, rapid setting admixtures, anti-freezing agents, and so forth. Those used in super-high-strength high performance concrete are mainly superplasticizers and retarders. An expansion admixture added in large amounts should be included in the binding material.

5.3.1 Superplasticizer

Before the emergence and application of water-reducing agents, in order to produce high strength or super-high-strength concretes, a stiff concrete with a low water/cement ratio and zero slump was used. The stiffness of concrete was increased by lowering the water/cement ratio; in order to form and compact such stiff concrete, a strong forming and compacting machine was needed resulting in large energy consumption, high noise, and intense work. In the 1950s and 1960s, in the former Soviet Union and in China, such concrete was developed. In 1959, China had produced a prestressed super-high-strength concrete frame, with a span of 18 m and C100 concrete.[39] Due to difficulties in forming and compacting, such concrete did not spread and develop further.

The invention and application of water-reducing agents (especially high-range water-reducing agents or superplasticizers) is a phenomenal progress in the development of concrete science and technology, which gives high fluidity to concrete mixes with moderate to low water/cement ratios, adjusted to the requirements of modern pumping technology of concrete.

The water-reducing rate of ordinary water-reducing agents is low and usually in range of 5% to 10%, whereas the water-reducing rate of a plasticizer is over 15%, and may be as high as 30% and over. The older water-reducing agents such as calcium lignosulphonate or molasses should not be used in high dosages (for example 0.3%), otherwise the setting of concrete will be retarded. A modern superplasticizer can be added in high dosage without any negative effect on the property of concrete.

At present, the plasticizers in widespread use in China are of two types, namely, naphthalene sulfonate formaldehyde condensate and melamine formaldehyde polymer.

5.3.1.1 Naphthalene sulphonate formaldehyde condensate

The naphthalene is sulfonated with oleum to produce naphthalene sulfonic acid after hydrolysis and reacted with formaldehyde to condense into a polymer; the residual sulfuric acid is neutralized with sodium hydroxide, filtered, and then dried. The reaction is as follows[40]:

1. Sulfonation of naphthalene

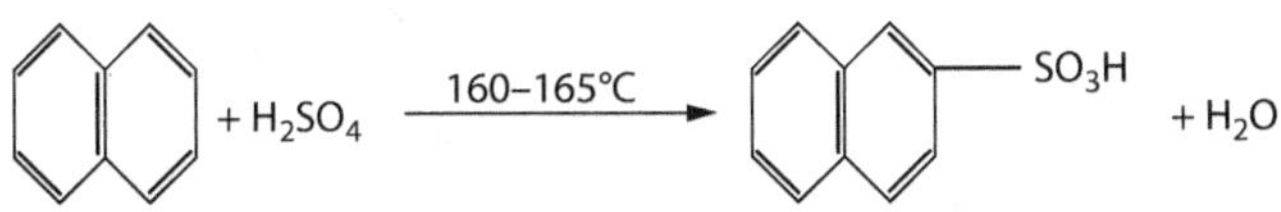

β sulfonic acid of naphthalene

A subordinate reaction during sulfonation gives α, sulfonic acid of naphthalene:

α - sulfonic acid of naphthalene

2. The hydrolysis of α, sulfonic acid of naphthalene gives

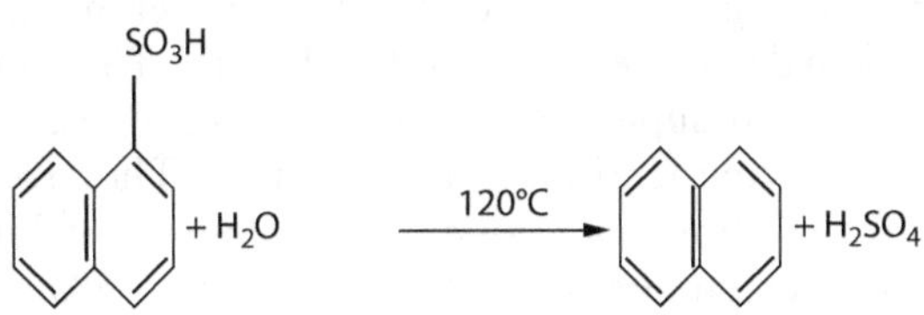

3. The condensation of β, sulfonic acid with formaldehyde leads to condensation:

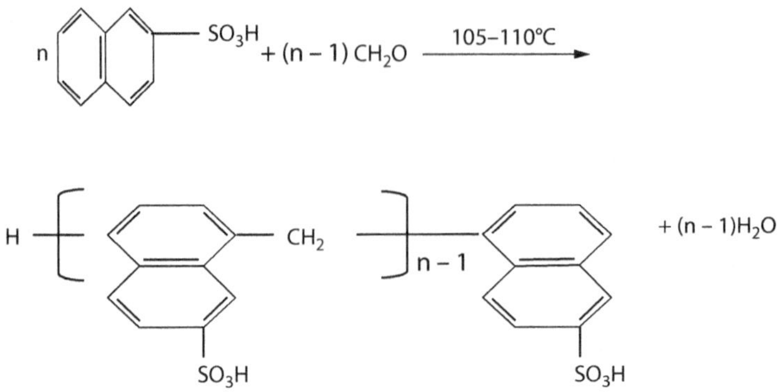

4. The neutralization by NaOH:

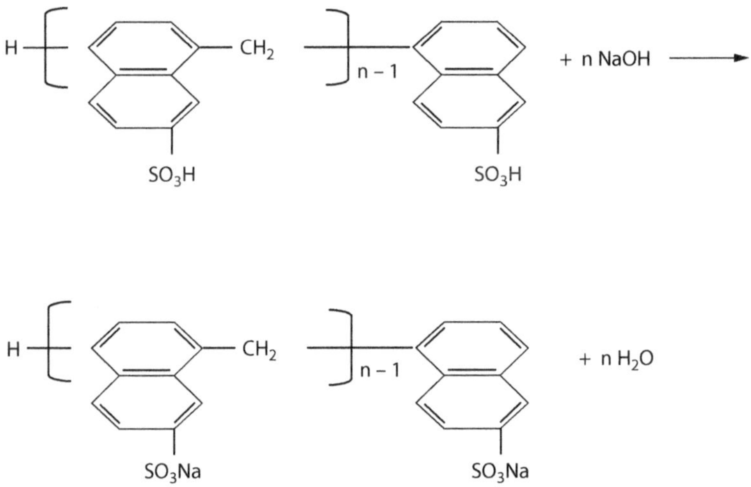

5. The residue sulphuric acid is neutralized by lime:

$$H_2SO_4 + Ca(OH)_2 \rightarrow CaSO_4 \downarrow + 2H_2O$$

The water-reducing rate of the superplasticizer naphthalene sulfonate formaldehyde condensate is related to the length of its molecular chain: the ideal superplasticizer is a long molecular chain with the degree of polymerization, n, more than 9. The product produced according to the above-mentioned technology is of high concentration; the mass percentage of Na_2SO_4 is less than 5%. If the neutralization is carried out by NaOH, more Na_2SO_4 is obtained (<25%), which is considered low concentration. The higher the concentration, the greater the water-reducing rate will be.

5.3.1.2 Melamine formaldehyde polymer

Melamine formaldehyde polymer is obtained by condensation of melamine, formaldehyde, and sodium bisulphate under certain conditions. The process can be divided into three stages[40]:

1. Synthesis of trimethylol melamine monomer:

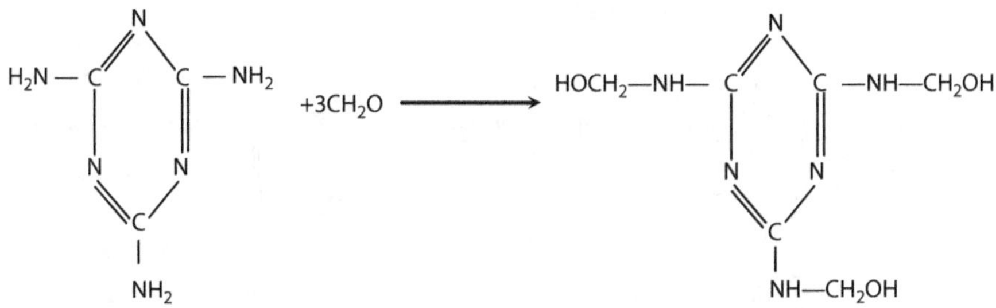

Melamine Trimethylol

   Three hydrogen molecules contained in the melamine molecule are able to react with formaldehyde to produce a methylol substitute; such reaction should proceed in a weak basic media (pH value 8.5).

2. Sulfonation of monomer:

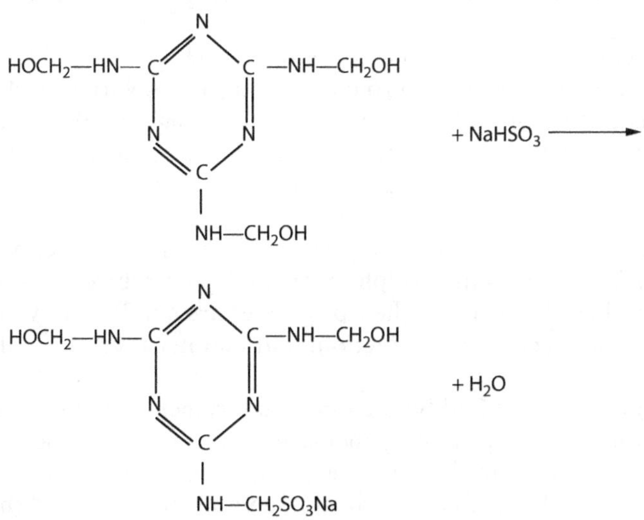

   To prepare naphthalene sulfonic acid, the typical sulphonator is concentrated sulphuric acid, whereas for sulphonation of trimethylol melamine, the sulphonator used is sodium bisulphate, sodium sulphite, or sodium pyrosulfite, and the reaction should proceed in a basic medium.

3. Condensation of monomers:

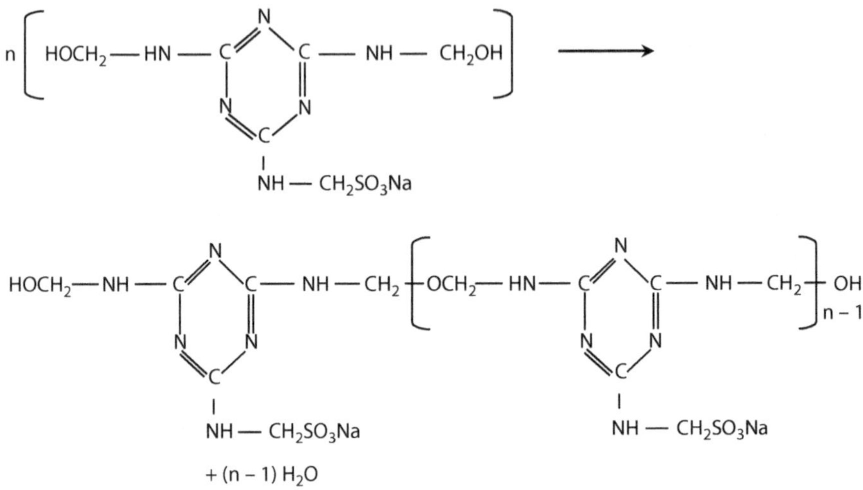

The condensation proceeds in a weak acidic media where the methylols are condensed into an ether linkage. The sulphonated methylol melamine monomers are linked into a linear resin by ether linkages and the relative molecule mass of this compound can reach 3000 to 30,000.

After completion of condensation, the pH is increased to 7 to 9 by dilute alkali solution to produce an aqueous solution with a solids concentration of 20%. By vacuum dewatering or pulverized drying, a white powder can be obtained. The water-reducing ability of such superplasticizers is very high with no air entrainment.

Along with the two common superplasticizers given above, new superplasticizers based on amino sulphonates and polycarboxylates are being produced and used in China. Their properties are excellent in view of dosage, dispersion effect, water reduction, and maintenance of the fluidity of the paste.[108]

For any given cement and binding material (cement + additive), the fluidity of the concrete is improved by increasing the dosage of superplasticizer. However, once a certain dosage is reached, there is little improvement in fluidity with a further increase of the dosage; this limit is called the *saturation point of the superplasticizer.* Addition of superplasticizer above the saturation point is ineffective and not economic, but it does not have a negative effect on the properties of the concrete, only a delay of setting time.

For super-high-strength high performance concrete, a superplasticizer with a water reducing rate more than 25% to 30% should be used, but as a pure product and not a combined mix of different products.

The addition method of superplasticizer has a significant influence on the fluidity of the concrete. After the raw materials and water have been mixed, the superplasticizer can be added with additional mixing to increase the fluidity of the concrete.

5.3.2 Retarders

In modern concrete technology, concrete is usually produced at ready-mix concrete plants, then transported by concrete truck to the building site. On site, the fresh concrete mix is pumped by a concrete pump through a pipe to the mold at casting point, where it is compacted often by vibration, and finished. The time for transportation depends on the distance from the plant to the site, the state of the road, and whether there are traffic hold-ups, and this time can vary from half an hour to 3 to 4 hours. However, the fresh concrete can set quickly, especially for high strength and super-high-strength concretes with low water/binder ratios and the addition of superplasticizers. The concrete mix can set during transport and lose its fluidity with a reduction of slump and even arrive in a state unsuitable for pumping or compaction. To avoid this situation, a retarder is added during the mixing of the concrete to delay the setting time of the concrete to allow successful implementation of the steps of transportation, pumping, casting, vibration, and finishing.

The setting time of cement is related to the hydration rate of the clinker minerals and condensation process of the cement/water colloid system along with the w/b ratio. Therefore, any admixture that can retard the hydration rate of the clinker phases and slow the condensation process of the cement/water colloid system can be used as a retarder. Generally speaking, any organic surface-active agent can be adsorbed onto the surface of a cement particle, preventing the reaction of the clinker phase and playing a set-retarding role. Some chemical compounds, such as gypsum, $CaSO_4 \cdot 2H_2O$, can react with hydration products of cement to form a complex salt, such as ettringite, which is precipitated on the surface of the cement particle, preventing the hydration of the cement and giving a retarding effect.

Hydroxycarboxylic acid molecules can be adsorbed onto C_3A in the cement, upsetting the formation of ettringite crystals and showing a retarding effect. When a phosphate dissolves in the water, it produces ions that are adsorbed by the cement particles to form a thin layer of Ca phosphate with low solubility, and this thin layer prevents the hydration of C_3A and crystallization of the ettringite; as a result, the setting of the cement is retarded.

The commonly used retarders are hydroxycarboxylic acids and their salts, for example, tartaric acid, citric acid, gluconic acid, salicylic acid; sugars and glucose, calcium sucrate, sodium gluconate; polyhydric alcohols; some inorganic substances such as boric acid and its salts, fluorosilicate, phosphonic acid and its salts (such as bi-polyphosphate, tri-polyphosphate, and multiple polyphosphate, in particular, sodium pyrophosphate $Na_4P_2O_7$), all of which are effective retarders.[40]

For some retarders, there are problems with matching them to superplasticizers and compatibility with the cement, so a test should be carried out before their use. The dosage should be strictly controlled to avoid excess prolongation of setting time and a negative effect on the early strength. Variation of the dosage depends on the type and content of cement, addition of mineral admixtures, and water content and temperature of the environment, all of which should be chosen by tests.

5.4 COARSE AGGREGATE

Aggregate is the basic component of concrete, occupying more than three fourths of the volume of the concrete, and sometimes called a filling material. In addition, it plays the role of providing a skeleton to support the pressure while the concrete is under load. In ordinary concrete (as opposed to lightweight concrete), it has greater density, strength, volume stability, and durability compared to the cement paste, which binds it together. From an economic aspect, it is much cheaper than cement, which is why concrete is a low cost building material. Usually, material with a particle size above 5 mm is called coarse aggregate, while the particles 0.15 to 5 mm are called fine aggregate or sand.

The coarse aggregate used for super-high-strength concrete should be chosen with the following characteristics: strength and elastic modulus, shape of the particle, surface character, maximum particle size, and so forth.

5.4.1 Strength and elastic modulus of coarse aggregate

In ordinary concrete, the overall strength of the concrete is much lower than that of the aggregate, so there has been no special interest in the strength of aggregate. In super-high-strength concrete, the strength of aggregate is a determining factor, because the strength of such concrete is higher than that of some rocks. To avoid failure of the aggregate, the strength of the aggregate should preferentially be higher than that of the concrete. Based on theoretical analysis,[41] taking into account the heterogeneity of a concrete structure, the maximum stress in an aggregate particle may reach 1.7 times that of the average stress within the concrete. Therefore, the strength of the parent rock of the coarse aggregate should preferably be more than 1.7 times that of the design strength of the concrete. Thus, dense, high strength natural rock suitable for crushing for coarse aggregate includes diabase, basalt, granite, quartzite, hard limestone, andesite, porphyrite, diorite, and gneiss. In addition, Sviridov[41] suggested that for concrete with a compressive strength of 150 MPa, the crushing value of the crushed stone should not exceed 6%, while for concrete with a compressive strength of 110 MPa the crushing value should not exceed 9%. The elastic modulus of the rock should also be high, because the elastic modulus of concrete depends to a large extent on the elastic modulus of the coarse aggregate.

5.4.2 Particle shape and surface state of coarse aggregate

The particle shape of the coarse aggregate determines its compaction. If a volume is filled with spherical particles of the same dimension, the percentage of the solids is 67% and the porosity 33%. To fill the same volume with particles of another shape in the same way, the volume the solid particles occupy is less than 67% of the volume and the porosity is greater than 33%. This difference in porosity is called the *angularity factor*; the greater the difference in porosity, the greater the angularity factor of the coarse aggregate, the worse the particle shape, and the more angular the particle. For commonly used aggregates, the angularity factor is 0 to 11.[42]

Another method to express the particle shape is by the ratio of surface area of the particle to its volume: the smaller the ratio, the closer the particle is to spherical, while the greater the ratio, the more the particle tends to be rectangular. Acicular and platey particles have a high surface area-to-volume ratio.

The shape of the coarse aggregate for super-high-strength concrete should be close to spherical with as much acicular and platey particles removed as possible, as they have a negative influence on the fluidity, strength, and durability of the concrete.

For a coarse aggregate, the surface of the particle should be rough to increase the adhesion of the aggregate with cement paste. Therefore, crushed stone should be chosen as the coarse aggregate for super-high-strength concrete, but not the gravel. In addition, the surface of the coarse aggregate should be clean and not contaminated with dust, silt, and so forth.

5.4.3 Maximum particle size

In a well-graded, coarse aggregate, the greater the maximum particle size, the smaller the total surface area will be, so the water requirement of the aggregate is lowered and the fluidity of the mix enhanced. However, the maximum aggregate size is restricted by the size of the elemental cross section, the degree of congestion of the reinforcement, and it cannot be enlarged arbitrarily; besides, an increase of aggregate particle size leads to more opportunity for segregation of the mix. In addition, some researchers believe that by allowing an increase of particle size of the coarse aggregate, there is more probability for textural defects such as microcracks within the particles generated during extraction (blasting) and processing (crushing), which will affect the strength of the concrete. Feng[4] and Sviridov et al.[41] suggest that the maximum particle size of coarse aggregate for high strength and super-high-strength concretes is preferably between 10 and 14 mm. However, during experiments on high strength and super-high-strength concretes, the author found that a coarser maximum particle size was satisfactory and coarse aggregate with a maximum particle size of 20 mm was used.

5.5 FINE AGGREGATE—SAND

According to the fineness modulus of sand M_x, given in Standard JGJ 52–92, the sand used for concrete can be divided into coarse sand (M_x = 3.7–3.1), medium sand (M_x = 3.0–2.3), and fine sand (M_x = 2.2–1.6). In addition, sometimes superfine sand (M_x = 1.5–0.7) and powder sand with $M_x < 0.7$ can be used beyond the standard.

Volkov[43] and the ACI 363 Committee[44] suggested that a coarse sand with a fineness modulus greater than 3.0 is the most suitable fine aggregate for the high strength and super-high-strength concretes. As a result, the water requirement of the concrete mix could be reduced, and the highest fluidity and strength could be obtained. The ACI 363 Committee[44] showed that in the case of sand with a fineness modulus less than 2.5, only a stiff mix could be obtained.

However, in China, the reserves of sand with high fineness modulus are limited and distributed unevenly, so for some areas, there is no coarse sand available and medium sand is limited as well. For example, in Chongqing and many other regions, only superfine sand and powder sand are available. Transporting medium or coarse sand long distances incurs a high transport fee (for example, medium sand delivered from Chengdu to Chongqing costs an extra RMB ¥100 per ton for transport). After about 50 years of extraction and utilization on a large scale, the original reserves of coarse and medium sand are almost exhausted. For many projects, the use of artificial sand obtained by crushing is normal; for example, in the large Three Gorges Project, mechanically processed granite sand was used in large amounts. Therefore, there is special significance in studying the preparation of high strength and super-high-strength concretes with lower quality fine sand, superfine sand, or even powder sand.

Since the 1950s, workers in China have studied the application of concrete made with superfine sand with great success and economic benefits have been gained. In 1965, the regulation for preparation and application of superfine sand concretes (BJG 19-65) was published; for concrete, it usually was no higher than C30 with a specified slump of no more than 3 cm. Nowadays, the C40 superfine sand concretes are widely used, and the application of C50 superfine sand, high strength pumping concrete has commenced.[45]

Is it possible to use superfine sand and powder sand for preparing super-high-strength high performance concretes? Before the study carried out by the author and his colleagues,[46] no work in this area had been reported, either in China or abroad.

The characteristics of superfine sand and powder sand differ from those of coarse and medium sand in the following ways. There is usually more silt content, with a poorer grading, larger porosity, and, in particular, the small particle size and large surface area lead to large water requirements for wetting the particle surface. Therefore, in designing the mix for concrete, if the same sand percentage is used for the fine and powder sand as for the coarse and medium sand, then to reach the same mix workability, the water content must be increased; at a constant cement content, the water/cement ratio increases and the strength of hardened concrete is decreased. If the water/cement ratio is kept constant, then the cement content must be increased raising the cost per unit volume of the concrete. Therefore, to prepare a concrete with superfine sand, the sand percentage should be chosen carefully taking into account the characteristics of the superfine sand.

In designing the correct concrete mix proportions within the macrostructure of the concrete, a single coarse aggregate particle should be separated by the mortar surrounding it; and within the mortar, a single sand particle should be separated by the cement paste surrounding it. The smaller the particle size of the sand, the closer the particles of coarse aggregate pack leaving less space for the mortar. To illustrate the relationship between the sand percentage to the particle size of the sand in concrete, Netkachev[47] suggested an ideal structural model of the concrete given in Figure 5.2.

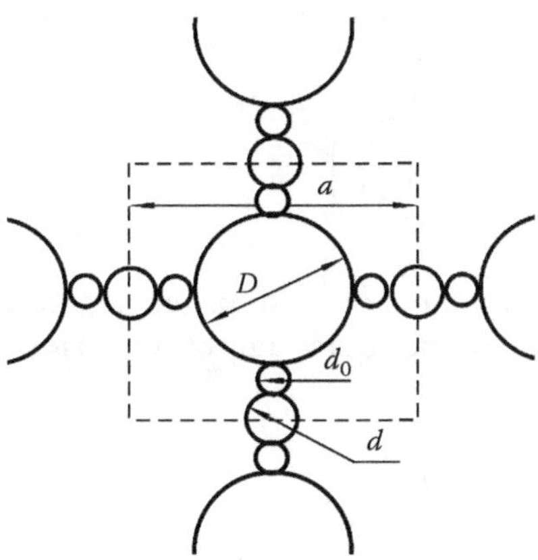

Figure 5.2 M.I. Netkachev Model.

It is supposed that all particles of the coarse aggregate and cement are spherical; between two particles of coarse aggregate there is only one sand particle and between the sand and coarse aggregate there is only one cement particle. The diameters of the coarse aggregate, sand, and cement particles are expressed by D, d, and d_o, respectively; and a, as length of the side of a unit cube confined by dotted lines in the model; therefore, the number of coarse aggregate particles in 1 m^3 of concrete, n_1, is

$$n_1 = \frac{1}{a^3} = \frac{1}{(D+d+2d_0)^3}$$

In 1 m^3 of concrete, the volume occupied by the coarse aggregate is

$$V_1 = n_1 \frac{1}{6}\pi D^3 = \frac{0.524D^3}{(D+d+2d_0)} = \frac{0.524}{\left(1+\frac{d}{D}+\frac{2d_0}{D}\right)^3} \tag{5.1}$$

In 1 m^3 of concrete, the volume occupied by the mortar is

$$V_{\text{mortar}} = 1 - V_1$$

Similarly, in 1 m^3 of mortar, the number of sand particles is

$$n_2 = \frac{1}{(d+d_0)^3}$$

Therefore, in 1 m^3 of concrete, the volume occupied by the sand is

$$V_2 = \frac{1}{6}\pi d^3 n_2 (1-V_1) = \frac{0.524}{\left(1+\frac{d_0}{d}\right)^3}(1-V_1)$$

If the relative densities of coarse and fine aggregate are the same, that is, $\gamma_1 = \gamma_2$, and the ratio, α, of the coarse aggregate mass (G_1) to that of the fine aggregate (G_2) is

$$\alpha = \frac{G_1}{G_2} = \frac{V_1\gamma_1}{V_2\gamma_2} = \frac{\left(1+\frac{d_0}{d}\right)^3}{\left(1+\frac{d}{D}+2\frac{d_0}{d}\right)^3(1-V_1)} \tag{5.2}$$

Then the sand percentage of the concrete, β, is

$$\beta = \frac{G_2}{G_1 + G_2} = \frac{1}{1+\alpha} \tag{5.3}$$

Suppose that the average particle size of the coarse aggregate in concrete is 10 mm, and the average particle size of the cement is 14 μm, with the average particle size of coarse sand, medium sand, fine sand, superfine sand, and powder sand 0.5, 0.4, 0.3, 0.2, and 0.1 mm, respectively; according to Equations (5.2) and (5.3), the calculated α and β values are given in Table 5.4.

It can be seen from Table 5.4 that as the particle size of the sand gradually decreases, the percentage of the sand filling the pores between the coarse aggregate is decreased gradually as well. Though the solid particles in concrete are not spherical and the particle filling is not as shown in Figure 5.2, the spacing between the coarse particles is influenced by the particle size of the sand; the smaller the particle size of the sand, the thinner is the mortar layer between the coarse aggregate particles. This is an undisputable fact and has been proved by experiment. Therefore, the calculated results given in Table 5.4 express the law of variation of sand percentage with the particle size of the sand.

A super-high-strength high performance concrete with a slump of 180 mm, and a compressive strength of 126.8 MPa at 28 days and 150.6 MPa at 365 days was prepared by the author and his colleagues using superfine sand from the Chongqing region (M_x = 1.2, silt content 0.95%). Likewise a super-high-strength high performance concrete with slump of 130 mm, compressive strength of 130.9 MPa at 28 days and strength of 151.2 MPa at 365 days (see Section 6.7) was made with powder sand (M_x = 0.63, silt content 4.65%) using Portland cement + an active mineral admixture and superplasticizer, and a low sand percentage, under normal conditions. This shows that by knowing the characteristics of the sand, it is possible to prepare super-high-strength high performance concrete with the normally neglected superfine or powder sands.

Whichever sand is chosen, the particles should be hard, and as Sviridov[48] suggested, sand used to produce 150-MPa super-high-strength concrete should have a breaking index of no more than 20%. In addition, its silt and mica contents should be as low as possible.

Table 5.4 Relationship between the sand percentage and the average particle size of the sand

Type of sand	*Coarse*	*Medium*	*Fine*	*Superfine*	*Powder*
Average particle size (cm)	0.05	0.04	0.03	0.02	0.01
Mass ratio of coarse to fine aggregate, α	1.671	1.831	1.983	2.289	2.876
Sand percentage, β (%)	37	35	33	30	26

Chapter 6

The problems with preparation techniques

6.1 TARGETS OF PREPARATION

Super-high-strength high performance concrete has a high load-bearing capacity with a strength–mass ratio that can reach 0.058 MPa/(kg·m^{-3}). This exceeds the specific strength of low carbon steel (Table 1.2), so it will find wide application in future engineering. In fact, low class super-high-strength high performance concrete has already found wide application in many modern civil engineering projects (see Chapter 11). To ensure increasing usage, therefore, it is very important to study and master the technology needed for its preparation.

The author has set criteria for its development.

1. Raw materials should be readily available and abundant. That is, ordinary raw materials must be used to develop such concrete for good economic benefit and the widespread use of raw materials.
2. General equipment and existing technology widespread within the construction industry should be used. This means regular batching and mixing, transport by concrete mixer truck, and placement by pumping or by bucket with compaction by vibration. Allowing preparation of such concrete to make use of modern concrete technology provides a reliable basis for its widespread application.

Certainly, the new material will have its own characteristics so there may be some improvement and enhancement of modern manufacturing and construction technology needed to adapt to these requirements. To conform to requirements of modern construction technology, the concrete should have high fluidity with good cohesion, slump of more than 180 mm, and low loss of slump over time.

Super-high-strength concrete as defined by the author includes concretes with strengths in the range of 100~150 MPa. It is relatively easy to produce super-high-strength concrete of 100 MPa; but to prepare super-high-strength concrete with strength around 150 MPa having high mix fluidity is very difficult. It requires careful selection of raw materials, and consideration of all factors that influence the parameters of the mix properties so that the given target can be reached.

As mentioned above, the durability index of concrete may be closely related to its strength, with the overriding factor for the relationship being porosity defined by water/cement (water/binder) ratio. It is clear that super-high-strength concrete is expected to have excellent durability. The author's target has been to prepare super-high-performance concrete using ordinary materials and general technology, which integrates good construction ability producing a concrete of super-high-strength and excellent durability.

Super-high-strength concrete has been defined by the author as a concrete with 90-day compressive strength of 100 mm × 100 mm × 100 mm cube specimens cured under standard conditions of more than 100 MPa. The reasoning for such a definition will be discussed and interpreted in Chapter 10.

6.2 RAW MATERIALS AND SAMPLE PREPARATION

6.2.1 Cement

Three cement types were utilized in an extended study of cement properties:

1. Type 525 OPC produced by the Chongqing Cement Plant, which has a measured 3-day flexural strength of 7.56 MPa, and a compressive strength of 31.6 MPa. The 28-day flexural strength of this cement is 9.4 MPa, and compressive strength is 54.0 MPa.
2. Type 525R cement produced by the Chongqing Tenghui Cement Plant. Its measured 28-day compressive strength is 65.6 MPa.
3. Type 625R cement produced by the Kunming Cement Plant. The measured 28-day compressive strength of this cement is 74.9 MPa.

The chemical compositions and specific surface areas are given in Table 6.1.

Table 6.1 The chemical composition and specific surface area of the cement

Type of cement	*Mass percentage of chemical composition*								*Specific surface area/ ($m^2\ kg^{-1}$)*
	CaO	*SiO_2*	*Al_2O_3*	*Fe_2O_3*	*TiO_2*	*MgO*	*SO_3*	*Loss of ignition*	
CQ525	62.00	20.50	5.90	3.10	0.60	1.80	2.17	0.80	330
TH525R	58.70	22.30	5.44	3.77	0.75	2.58	2.67	1.60	
KM625R	62.60	19.90	5.90	3.90	0.42	1.60	2.80	0.66	380.6

6.2.2 Active mineral admixtures

1. Silica fume: two kinds of silica fume were used, one from the Electric Furnace Steel Plant of Tangshan Steel & Iron Co. and the other from the Guiyang Qinzhen Iron Alloy Factory.
2. Ground granulated blast furnace slag from the Chongqing Steel & Iron Co.
3. A Chinese Class II fly ash from the Chongqing Power Plant.
4. Lithium slag, from Sichuan Santai Lithium Industry Co.
5. X powder and Y powder.

Except for the silica fume, the other mineral admixtures were ground in a 200-L vibrating mill after drying. The chemical compositions and specific surface areas are given in Table 6.2.

It can be seen from Table 6.2 that for both silica fumes, the mass percentages of SiO_2 are >90% with specific surface areas of about 20m²/g, making them suitable for use without further treatment. They serve as the basic additives for preparation of super-high-strength high performance concrete. The alkali factor of the slag is 1.02 making it basically neutral. Although the fly ash is classified as a Class II commercial ash, its loss of ignition at >8.0% means it does not meet the specification for a Class II fly ash. The lithium slag is a fine powder and with its specific surface area before grinding of 319 m^2/kg, is close to that of the cement. It has good grindability and its specific surface area can reach 949.3 m^2/kg after grinding in a vibrating mill for 15 min, 1126.8 m^2/kg after 30 min grinding, and 1379.6 m^2/kg after 60 min grinding. Its high mass percentage of SiO_2 reaching 74.57% gives it good pozzolanic activity.

Table 6.2 Mass percentage of chemical composition and specific surface area of active mineral admixtures

	Mass percentage of chemical composition								
Additive	*CaO*	*SiO_2*	*Al_2O_3*	*Fe_2O_3*	*TiO_2*	*MgO*	*SO_3*	*Loss of ignition*	*Specific area/ (m^2g^{-1})*
TS silica fume	0.45	91.27	0.17	0.45		0.92		2.88	~200
QZ silica fume	0.56	94.90	0.49	1.07		0.70		1.90	~200
CQ slag	40.00	33.14	12.91	2.80	2.08	6.75			8.8
CQPL fly ash	3.40	40.00	25.30	15.40	3.20	0.49	0.7	8.50	7.017
Li slag	3.32	63.20	21.00	0.98	0.15	0.39	3.44	5.32	9.493
X powder	1.49	74.57	11.69	1.25	0.13	0.27	0.075		17.470
Y powder	10.80	39.10	15.85	13.92	2.60	0.95		13.15	8.813

6.2.3 Coarse aggregate

Two kinds of coarse aggregates were used:

1. Crushed basalt from Puge County, Xichang Region of Sichuan Province: the strength of the mother rock was 230 MPa (for a ϕ50 mm × 50 mm specimen) with a crushing index of 3.7%.
2 Crushed limestone from the Xiaoquan region, Chongqing City: the strength of the mother rock was 154.7 MPa, the crushing index was 10.3%, the maximum particle size was 20 mm except specified otherwise.

6.2.4 Fine aggregate

1. Medium sand from Jianyang, Sichuan Province with a fineness modulus of 2.40.
2. Superfine sand from the Qujiang River, Hechuan city in Chongqing metropolitan, with a fineness modulus of 1.20.
3. Powder sand from the Jialin River in Chongqing with a fineness modulus of 0.63.
4. Mixed sand: 2/3 prepared sand from <5 mm crushed limestone and 1/3 Jianyang medium sand, with a fineness modulus of 3.5.

6.2.5 Water-reducing admixtures (WRA)

1. Water-reducing agent of a lignosulfonate type.
2. Superplasticizer of naphthalene sulfonate formaldehyde polymer (S-20).
3. Superplasticizer of a naphthalene sulfonate, retarding type.
4. Superplasticizer of melamine formaldehyde polymer (DSF-2).

The dosage of water-reducing admixture is taken as the volume in liters per m^3 concrete or the mass percent of binder (cement + active mineral admixture). In the case of a solid, the conversion is straightforward. The formulation was mixed in either a large drum mixer (375 L) or in a small one (30 L).

The mass was processed through weighing, dry mixing, wet mixing, fluidity measurement, compaction by vibration, standard curing up to specified age (3 d, 28 d, 56 d, 90 d, or 365 d), and strength testing. The dimensions of the specimens prepared were as follows: for compressive and splitting tensile tests, 100 mm × 100 mm × 100 mm cubes; for axial compression and elastic modulus tests, 100 mm × 100 mm × 300 mm prisms; for the flexural bending test, 100 mm × 100 mm × 400 mm beams; and for the bond to steel bar, specimens were 100 mm × 100 mm × 200 mm.

The author and his colleagues have studied the influence of cement grade, binder content, water/binder ratio, dosage of water-reducing admixture, maximum particle size of the coarse aggregate, type of coarse aggregate, type of fine aggregate, sand percentage, type and dosage of mineral admixture, curing age, and curing conditions on the fluidity and strength of super-high-strength high performance concrete made with the materials mentioned above, the details of which are given in the following sections.

6.3 INFLUENCE OF CEMENT GRADE AND BINDER CONTENT ON THE FLUIDITY AND STRENGTH OF CONCRETE

Test results of concrete made with Chongqing 525R and Kunming 625R Portland cement are given in Table 6.3. It can be seen from this table that the grade and fineness of the cement have a significant influence on the fluidity and strength of the concrete. The concrete made with cement grade 625R had a higher strength than that made with cement 525R (comparison of No. 1 with No. 3 and No. 2 with No. 4). This is to be expected, however, the increase in concrete strength was not directly proportional to the fineness of the cement. The slump of two concrete formulations was about the same, but the spread of concrete from type 625R cement was much lower.

Special attention should be paid to the mix formulations No. 2 and No. 4 in Table 6.3, in which the binder content was 650 kg/m^3. The water/binder

Table 6.3 Influence of cement grade on the fluidity and strength of the concrete

No.		*1*	*2*	*3*	*4*
Cement grade		525R	525R	625R	625R
28 d compressive strength of cement/MPa		54.0	54.0	74.9	74.9
Silica fume content/%		10	10	10	10
Slag content/%		—	30	—	30
Total binder content (kg/m^3)		600	650	600	650
Medium sand percentage/%		40	40	40	40
Water binder ratio		0.25	0.20	0.25	0.20
Water reducing agent/(L·m^{-3})		21	31	21	31
Fluidity	Slump	235	255	234	230
	Spread	586	570	430	420
28 d apparent density of concrete (kg/m^3)		2550	2554	2557	2558
Compressive strength/MPa	3 d	68.7	65.0	64.7	71.5
	28 d	105.2	109.8	117.3	121.5
	56 d	112.0	121.2	117.3	121.5
	90 d	126.0	131.6	126.7	139.9
	365 d	126.8	139.1	144.1	153.0

Note: Naphthalene series water-reducing agent S-20 (relative density 1.2, solid content 40%).

ratio was reduced to 0.20, and 30% more slag was added to the binder, so that the total content of the mineral admixture was 40%. The cement content was in fact only 390 kg/m^3 compared with 540 kg/m^3 (comparing No. 2 with No. 1 and No. 4 with No. 3), but the slump was more than 230 mm. Except for the 3 d strength of mix No. 2, the strength at all ages are much higher, especially at later ages. Because of the low content of the cement in these two mixes, the heat of hydration of the concrete could be much lower.

The influence of binder content on the fluidity and compressive strength of super-high-strength high performance concrete was studied with test results for 550, 600, and 650 kg/m^3 cement contents given in Table 6.4 and Figure 6.1. It can be seen that with an increase in binder content, the fluidity of the concrete mix increased greatly. Because the w/b ratio is kept constant, this leads to an increase in the amount of cement paste with a corresponding decrease in the amount of aggregate. The binder content also has an influence on the compressive strength of the concrete. Decreasing the binder content gave rise to low strength due to poor fluidity and the concrete was difficult to place and compact. In addition, the strength of concrete made with medium sand was higher than that of concrete made with superfine sand.

Table 6.4 Influence of binder content on the fluidity and strength of concrete

No.		*1*	*2*	*3*	*4*	*5*	*6*
Binder content (kg/m^3)		550	600	650	550	600	650
Silica fume/%		10	10	10	10	10	10
Type of sand		Medium	Medium	Medium	Superfine	Superfine	Superfine
Sand percentage/%		40	40	40	25	25	25
Water binder ratio		0.20	0.20	0.20	0.20	0.20	0.20
Water reducing agent/ (L·m^{-3})		19	21	31	19	21	31
Fluidity	Slump/mm	8	55	193	20	60	180
	Spread/mm	—	—	310	—	—	320
28 d apparent density of concrete (kg/m^3)		2578	2553	2576	2556	2578	2559
Compressive strength/ MPa	3 d	—	—	88.2	89.4	94.0	85.0
	28 d	120.9	118.4	131.4	116.0	126.6	126.8
	56 d	—	—	137.0	123.4	132.2	132.2
	90 d	130.1	134.6	143.4	128.9	134.4	134.7
	365 d			156.6	151.6	154.3	150.6

Notes: (1) Cement type 625R was used throughout. (2) Due to the lack of specimens for mix No. 1 and No. 2, only the 28 d and 90 d strength are given. (3) Water-reducing agent was a Naphthalene series S-20.

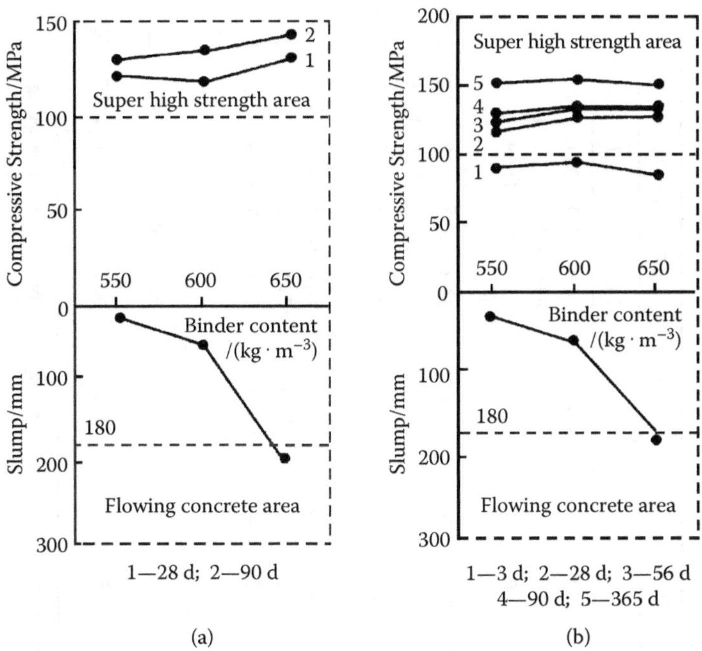

Figure 6.1 Influence of binder content on the fluidity and strength of the concrete: (a) medium sand; (b) superfine sand.

6.4 INFLUENCE OF WATER/BINDER RATIO ON THE FLUIDITY AND STRENGTH OF CONCRETE

It is well known that it is the water/binder ratio that is the basic factor affecting the strength of concrete. Provided the concrete is sufficiently fluid so the mix can be compacted fully by the formwork compacting equipment, the lower the water/cement ratio, the higher the strength of the concrete—often called the water/cement ratio rule. This rule also applies to high strength and super-high-strength concrete as well, because the lower the water/binder ratio, the lower the porosity of the hardened concrete. Therefore, to prepare super-high-strength high performance concrete, we used a water/binder ratio of ≤0.3, viz., 0.20, 0.25, and 0.30. In one case, 0.18 was used. The cements used were Chongqing 525R cement and Kunming 625R cement. Test results are given in Table 6.5 and illustrated in Figure 6.2.

It can be seen from Figure 6.2 that the fluidity of the three groups of mixes decreased with a reduction of water/binder ratio. This is due to the fact that with the same binder content, reducing the water/binder ratio

Table 6.5 Influence of water/binder ratio on the fluidity and strength of the concrete

Group	No.	Cement type	Binder content ($kg{\cdot}m^{-3}$)	Silica fume dosage (%)	Type of sand	Sand percentage (%)	Water/binder ratio	WRA (Lm^{-3})	Fluidity (mm)		28 d apparent density of concrete ($kg{\cdot}m^{-3}$)	Compressive strength (MPa)				
									Slump	Spread		3 d	28 d	56 d	90 d	365 d
A	1	Chongqing 525R	600	10	Medium	40	0.30	17	280	800	2497	54.8	92.0	106.1	104.1	116.3
	2		600			40	0.25	21	260	670	2518	66.6	105.0	112.0	114.1	120.6
	3		600			40	0.20	29	110	—	2524	73.1	111.9	118.9	116.5	122.5
B	4	Kunming 625R	550	10	Medium	25	0.30	19	235	570	2513	56.3	99.3	99.8	110.8	122.8
	5		550			25	0.25	19	190	415	2574	71.5	100.1	121.5	121.5	127.3
	6		550			25	0.20	19	30	—	2566	93.7	120.9	130.5	137.2	155.6
C	7	Kunming 625R	550	10	Mixed	40	0.30	19	195	400	2572	56.2	100.7	103.4	105.0	117.3
	8		550			40	0.25	19	150	350	2611	75.1	112.3	120.2	115.8	134.2
	9		550			40	0.20	19	5	—	2613	95.7	121.3	129.5	129.7	157.7
D	10	Kunming 625R	650	10	Mixed	25	0.18	31	10	—	2611	95.9	140.2	139.1	142.7	155.7
E	11	Kunming 625R	650	10	Medium	25	0.18	31	9	—	2643	100.3	125.5	135.4	138.7	154.5

Notes: Mix No. 1, No. 2, and No. 3 were cured under normal conditions, then removed to the roof for natural curing. In mix No. 11, the coarse aggregate used was crushed basalt. Water-reducing agent was naphthalene series S-20.

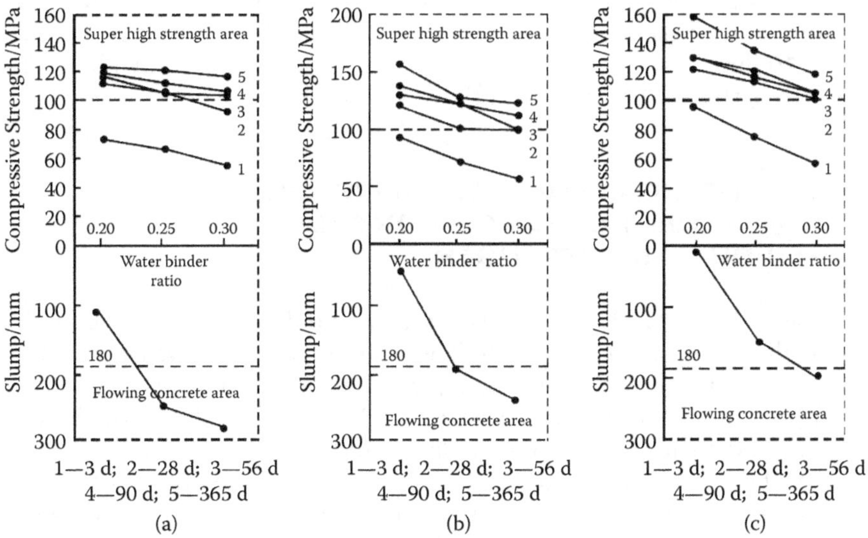

Figure 6.2 **Influence of water binder ratio on the fluidity and strength of the concrete: (a) 525 cement, binder content 600 kg/m^3, medium sand percentage 40%; (b) 625 cement, binder content 550 kg/m^3, medium sand percentage 25%; (c) 625 cement, binder content 550 kg/m^3, mixed sand percentage 40%.**

leads to a reduction of water content. When the water/binder ratio was >0.25, concrete met the requirements for flowing concrete (slump > 180 mm). For Group A mixes, when the binder content and sand percentage were high, the fluidity of the concrete mix was good, particularly when the water/binder ratio was 0.3. For this sample, the slump of the mix reached 280 mm and the spread 800 mm, an extremely high value. The maximum particle size of the coarse aggregate used in our tests was 20 mm. The concrete mix cone was 300 mm high collapsing by 280 mm to form a round cake with a diameter 800 mm and thickness 20 mm, both in the center and at the edge. This thickness was just the maximum particle size of coarse aggregate. Such a large slump and large spread have not been reported previously, both at home and abroad. It is understandable that to achieve such a high fluidity at a fixed binder content, when the water/binder ratio was increased, the water content also increased. Though the water/binder ratio was rather high, the 28-day strength still reached 92 MPa, with a 56-day strength of 106 MPa and a 365-day strength of 116 MPa.

For Group B mixes, the binder content was smaller and the sand percentage lower, so the fluidity of the mixes was also lower. However, because the specific surface of the cement was higher, the strength of concrete was higher, a strength of 155.6 MPa was reached at the age of 365 d with a water/binder ratio of 0.20. For groups D and E, the water/binder ratio was

0.18, and although the binder content was high, the water/binder ratio was too low, so with the lower sand percentage, the fluidity of the mix was too low at 10 mm and 9 mm to allow adequate compaction, although both the early strength and the later strength were very high.

6.5 THE INFLUENCE OF WATER-REDUCING ADMIXTURE DOSAGE ON CONCRETE FLUIDITY AND STRENGTH

In these tests, the 625R cement was used at a binder content of 600 kg/m^3 and a silica fume content of 10%. The water/binder ratio was set at 0.25 and the water-reducing admixture (naphthalene sulphonate superplasticizer) used at the dosage of 0.8%, 1.1%, 1.4%, and 1.7% of the binder content. Test results are given in Table 6.6 and shown in Figure 6.3.

It can be seen from Figure 6.3 that by increasing the dosage of water-reducing admixture, the fluidity of the concrete was enhanced up to 1.1 wt% of cement, which is the saturation point of this water-reducing admixture, so when the dosage exceeded 1.4%, the fluidity decreased due to segregation of the mix. The water-reducing admixture dosage has a rather large influence on the early strength at 3 d with low dose rates giving higher early strengths, but it has less influence on the 28-day or later strengths.

Table 6.6 Influence of WRA dosage on the fluidity and strength of super-high-strength high performance concrete

No.		*1*	*2*	*3*	*4*
Grade of cement		625R	625R	625R	625R
Binder content (kg/m^3)		600	600	600	600
Silica fume content/%		10	10	10	10
Medium sand percentage/%		40	40	40	40
WRA dosage/%		0.8	1.1	1.4	1.7
Water binder ratio		0.25	0.25	0.25	0.25
Fluidity/mm	Slump	130	235	250	234
	spread	250	440	526	435
28 d apparent density of concrete (kg/m^3)		2542	2552	2561	2557
Compressive strength/MPa	3 d	82.6	79.0	76.8	64.7
	28 d	122.6	118.0	113.7	117.3
	56 d	129.0	131.3	128.7	129.4
	90 d	137.1	137.1	132.4	126.7
	365 d	137.7	140.0	142.0	144.3

Note: Naphthalene series water-reducing agent S-20.

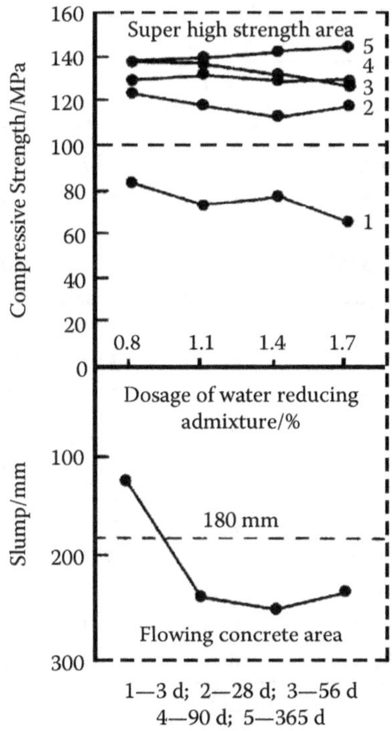

Figure 6.3 Influence of dosage of water-reducing admixture on the fluidity and strength of the concrete.

6.6 THE INFLUENCE OF COARSE AGGREGATE TYPE AND MAXIMUM PARTICLE SIZE ON THE FLUIDITY AND STRENGTH OF CONCRETE

Two types of coarse aggregates were chosen for the study, namely, crushed limestone and crushed basalt. The cements used were 525R and 625R at a binder content of 650 kg/m^3, water/binder ratio 0.2. Details of the mix proportions and test results are given in Table 6.7. It can be seen from this table that when basalt was used, the mix had better fluidity. Although the strength of the basalt aggregate was very high, concrete made with crushed basalt did not have any advantage in compressive strength.

Three sizes of crushed limestone were used, passing 30, 20, and 10 mm. With crushed stone of 30 mm, equal proportions of 5~10 mm grading, 10~20 grading, and 20~30 mm grading were blended; with a maximum particle size of 20 mm, the 5~10 mm grading and the 10~20 grading were used in a ratio of 1:2. The cement grade used was 625R with a binder content of 650 kg/m^3, a silica fume addition of 10%, a medium sand content of

Table 6.7 Influence of the type of coarse aggregate on the fluidity of the mix and strength of the concrete

No.		*1*	*2*	*3*	*4*
Grade of cement		525R	525R	625R	625R
Binder content (kg/m³)		650	650	650	650
Silica fume content/%		10	10	10	10
Slag content/%		30	30	—	—
Medium sand percentage/%		40	40	40	40
Kind of coarse aggregate		Limestone	Basalt	Limestone	Basalt
Water binder ratio		0.20	0.20	0.20	0.20
WRA dosage/(L·m⁻³)		31	31	31	31
Fluidity/mm	Slump	255	260	193	205
	spread	570	610	310	360
Apparent density of concrete at 28 d (kg/m³)		2554	2561	2576	2603
Compressive strength/MPa	3 d	65.0	59.4	88.2	93.7
	28 d	109.8	107.5	131.4	127.0
	56 d	121.2	120.1	137.0	140.1
	90 d	131.6	120.1	137.0	140.1
	365 d	130.8	133.3	156.6	160.7

Note: Naphthalene series water-reducing agent S-20.

40%, and a water/binder ratio of 0.20. Test results are given in Table 6.8 and shown in Figure 6.4. It can be seen from Figure 6.4 that the fluidity of the mix was enhanced with an increased maximum particle size in the coarse aggregate. This is because increasing the maximum particle size decreased the total surface area of coarse aggregate so that the water required for wetting was reduced. The maximum particle size of coarse aggregate has almost no influence on the strength of concrete as the strengths at different ages showed little difference for the three mix formulations. Gjorv[13] suggested that the maximum particle size of coarse aggregate for super-high-strength concrete should not exceed 14 mm, because the larger the coarse aggregate, the more possibility there was for introducing microcracking during its processing, which leads to lowering of the concrete strength. However, our study did not reach the same conclusion.

6.7 THE INFLUENCE OF SAND TYPE AND PERCENTAGE AS FINE AGGREGATE ON THE FLUIDITY AND STRENGTH OF CONCRETE

In our study, 625R Portland cement from Kunming Cement Plant and 10% addition of silica fume from Tangshan city together with crushed limestone

Table 6.8 Influence of maximum particle size of aggregate on the fluidity of the concrete mix and strength of the concrete

No.		*1*	*2*	*3*
Grade of cement		625R	625R	625R
Binder content (kg/m^3)		650	650	650
Silica fume content/%		10	10	10
Medium sand percentage/%		40	40	40
WRA dosage/%		31	31	31
Maximum particle size of aggregate/mm		30	20	10
Water binder ratio		0.20	0.20	0.20
Fluidity/mm	Slump	215	193	90
	spread	370	310	230
28 d apparent density of concrete (kg/m^3)		2572	2576	2569
Compressive strength/MPa	3 d	90.8	88.2	86.5
	28 d	132.1	131.4	136.7
	56 d	137.1	137.0	136.8
	90 d	140.1	143.1	142.9
	365 d	156.9	156.6	156.7

Note: Naphthalene series water-reducing agent S-20.

(maximum particle size of 20 mm) were used with a naphthalene sulphonate formaldehyde polymer superplasticizer added at a fixed percentage of binder content.

In this study, mixed sand (XS, coarse sand), medium sand (MS), superfine sand (SFS), and powder sand (PS) were selected. The choice of a mixed sand was due to the rather low fineness modulus of the medium sand (2.4) used in the test, while the fineness modulus for the crushed machined sand was high (4.4), but it was deficient in fine particles and so not suitable for use alone. Mixing crushed sand with natural medium sand in a ratio of 2:1 gave a fineness modulus for mixed sand of 3.5 and it is classified as coarse sand. In the test, the water/binder ratio ranged from 0.30 to 0.20, with the binder content varying from 550 to 650 kg/m^3. The test results are given in Table 6.9.

By comparing mix No. 1 with mix No. 2, it can be seen that although the fineness modulus of the mixed sand was high, due to the angular form of the crushed sand particles with rough surfaces, the fluidity of the concrete mix was lowered. Despite this, the strengths at different ages for both mixes were similar, so in regions where coarse or medium sands are not available, an artificial crushed sand with the proper grading could be used to reduce the cost of concrete.

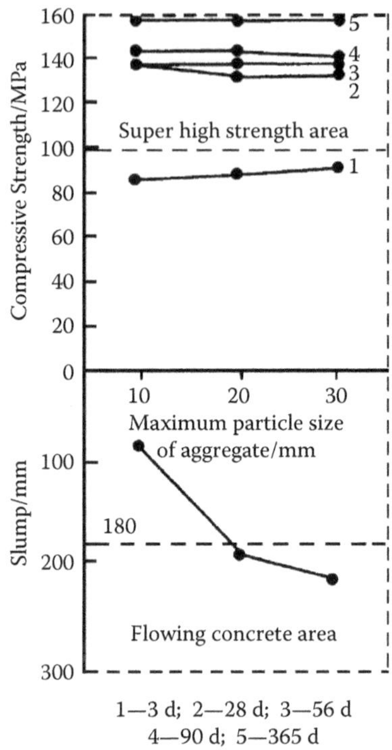

Figure 6.4 Influence of maximum particle size of aggregate on the strength and fluidity of the concrete.

With a binder content of 650 kg/m, sand percentage of 40%, and a water/binder ratio of 0.20 (mix No. 2), a super-high-strength concrete with a slump > 180 mm and reasonable strength at all ages can be obtained. Using the same mix proportions, an excellent super-high-strength concrete can be produced by only changing the sand to 25% superfine sand (mix No. 10) or 21% powder sand (mix No. 23). However, the concrete mix slump was lower than that for a medium sand concrete mix (193 mm for medium sand concrete, 180 mm for superfine sand concrete, and 130 mm for powder sand concrete), although the strength at all ages was lower than that of medium sand concrete. Therefore, by using superfine sand or powder sand, and regulating the parameters for mix proportion, a super-high-strength concrete with good fluidity can be prepared (see Figure 6.5).

Based on the data in Table 6.9, a relationship between fluidity and strength of superfine sand and powder sand concretes with different water/binder ratios can be drawn. It can be seen that the lower the water/binder ratio, the lower the fluidity of the mix, although the concrete strengths at all ages were higher. Certainly, for different water/binder ratios, the fluidity of powder sand concrete is worse than that containing superfine sand.

Table 6.9 Parameters of mix proportion, fluidity, and compressive strength of mixed sand, medium sand, superfine sand, and ultrafine sand super-high-strength high performance concretes

No.[a,b]	B (kg/m³)	Kind of sand	S/%	W/B	WRA (L/m³)	Fluidity/mm		ρ_a at 28 d kg/m³	Compressive strength/MPa				
						Slump	Spread		3 d	28 d	56 d	90 d	365 d
1	550	XS	40	0.20	19	5	—	2613	95.7	121.3	129.5	129.7	157.7
2	550	MS	40	0.20	19	11	—	2578		120.9		130.1	
3	650	MS	40	0.20	31	193	310	2576	88.2	131.4	137.0	143.1	156.6
4	600	MS	45	0.25	21	235	445	2578	70.9	112.5	120.0	128.1	136.1
5	600	MS	40	0.25	21	234	435	2557	64.7	117.3	129.4	126.7	144.1
6	600	MS	35	0.25	21	220	445	2581	72.5	111.2	120.4	124.0	137.3
7	600	MS	25	0.25	19	210	440	2592	69.7	105.7	109.9	117.0	127.3
8	550	MS	25	0.20	19	30	—	2566	93.4	120.9	130.5	137.2	155.6
9	550	MS	25	0.30	19	235	570	2513	56.3	99.3	99.8	110.8	122.8
10	650	SFS	25	0.20	31	180	320	2559	85.0	126.8	132.2	134.7	150.6
11	600	SFS	25	0.20	21	60	—	2578	94.1	126.6	132.2	134.4	154.3
12	600	SFS	25	0.25	21	255	570	2559	66.1	104.0	110.5	121.9	133.3
13	600	SFS	25	0.30	17	260	690	2501	51.3	97.4	104.0	104.5	113.9
14	550	SFS	25	0.20	19	20	—	2558	89.4	116.6	123.4	128.9	151.6
15	550	SFS	25	0.25	19	200	440	2554	71.0	106.3	115.6	118.4	131.5
16	550	SFS	25	0.30	19	240	615	2461	53.2	100.1	97.4	107.3	113.1
17	600	PS	21	0.20	26	10	—	2585	96.3	142.1	134.9	150.7	151.6
18	600	PS	21	0.25	26	205	485	2548	71.7	115.3	121.3	127.8	127.9
19	600	PS	21	0.30	26	223	635	2490	47.1	84.2	92.7	95.2	97.0

continued

Table 6.9 (continued) Parameters of mix proportion, fluidity, and compressive strength of mixed sand, medium sand, superfine sand, and ultrafine sand super-high-strength high performance concretes

No.[a,b]	B (kg/m^3)	Kind of sand	S/%	W/B	WRA (L/m^3)	Fluidity/mm		ρ_a at 28 d kg/m^3	Compressive strength/MPa				
						Slump	Spread		3 d	28 d	56 d	90 d	365 d
20	600	PS	17	0.25	26	185	533	2554	68.2	106.7	110.2	113.0	116.1
21	600	PS	25	0.25	26	215	503	2532	70.4	116.6	119.1	124.6	126.5
22	600	PS	29	0.25	26	190	—	2548	67.7	110.1	122.8	124.5	128.8
23	650	PS	21	0.20	28	130	—	2582	84.1	130.9	136.5	139.6	151.2
24	650	PS	21	0.25	28	230	628	2533	68.1	111.1	111.7	118.4	121.8
25	550	PS	21	0.25	24	185	388	2488	74.1	111.9	119.3	117.7	121.1

Notes: B = binder content; S = sand percentage; W/B = water/binder ratio; WRA = dosage of water-reducing admixture; 28 dρ_a = 28-day apparent density of the concrete.

[a] For mix No. 1 through No. 16, the water-reducing admixture was naphthalene series S-20.

[b] For mix No. 17 through No. 26, melamine water-reducing admixture DFS-2 (relative density 1.125, solid content 31%) was used.

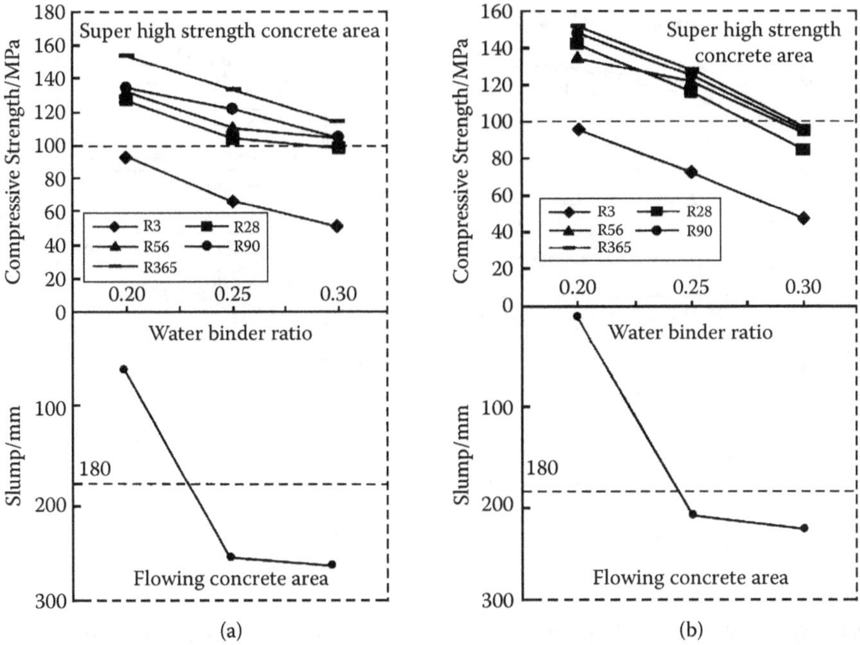

Figure 6.5 Relationship between the strength and fluidity of the concrete with water/binder ratio (binder content 600 kg/m^3): (a) superfine sand concrete (sand percentage 25%); (b) powder sand concrete (sand percentage 21%).

It can be seen from Figure 6.6(a) that when binder content was 600 kg/m^3, water/binder ratio was 0.25, and sand percentage was 40%, the concrete had the highest fluidity and strength (except at age of 3 d). However, with different sand percentages, the differences in slump and spread were not large. With a sand percentage of 25%, the concrete had minimum fluidity and strength, so the sand content was not enough.

From Figure 6.6(b), it can be seen that by increasing the percentage of powder sand concrete from 17% to 29%, the fluidity and strength of the concrete varied, with the optimum percentage being between 21%~25%, when both the fluidity and the strength reached a maximum.

6.8 THE INFLUENCE OF ACTIVE MINERAL ADMIXTURE DOSAGE AND TYPE ON THE FLUIDITY AND STRENGTH OF CONCRETE

In Section 4.4, the influence of silica fume, slag, fly ash, X powder, and Y powder on the fluidity and strength of concrete was discussed. The material

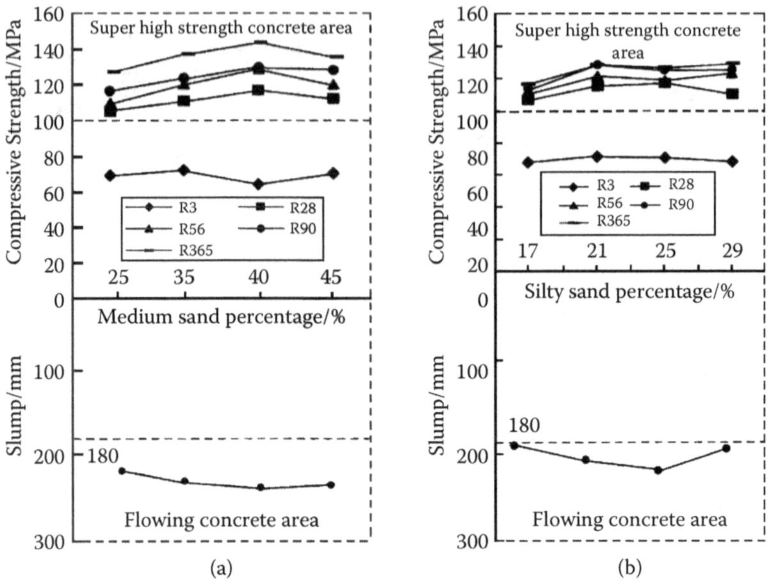

Figure 6.6 Influence of sand percentage on the fluidity and strength of super-high-strength high performance concrete (binder content 600 kg/m^3, water/binder ratio 0.25): (a) medium sand concrete; (b) powder sand concrete.

with the most influence on fluidity and strength was silica fume followed by X powder. From Table 6.2, it can be seen that the contribution silica fume makes to high strength and fluidity is due to its high mass percentage of available SiO_2 (>90%) and high surface area (about 20 m^2/g) and small particle size (0.1~0.2 μm) making it very reactive. This shows, therefore, that for a mineral admixture to have high pozzolanic activity, it should have (1) a large mass percentage of amorphous readily available SiO2; (2) fine particles; and (3) hard grains with a spherical shape and smooth surfaces. Silica fume has basically all three of these characteristics whereas fly ash only has the third one. Although blast furnace slag does not have these characteristics, it has a large mass percentage of CaO with potential hydraulicity, which can be mobilized, so in comparison with other additives, it can be used at high addition rates to provide high strength.

The author has studied the possibility of using a lithium slag as an active mineral admixture for super-high-strength high performance concrete.[49] Test results showed that with ground lithium slag added at levels between 10% and 30%, a super-high-performance concrete with good fluidity could be obtained. This is due to the high mass percentage of SiO_2 and Al_2O_3 in its chemical composition and rather high portion of glass making its activity index higher than that of fly ash.

The author has also studied the influence of silica fume addition on the fluidity and strength of super-high-strength high performance concrete. In that test, the materials used were 525R OPC from Chongqing Tenghui Cement Co, silica fume from Guizhou Qinzhen County, limestone from Chongqing Xiaoquan (max size 20 mm), medium sand, and a melamine-based superplasticizer DSF-2. The mix proportions and test results are given in Table 6.10 and shown in Figure 6.7.

It can be seen from Table 6.10 and Figure 6.7 that without silica fume addition, the slump of the mix was 0. Adding 5% of silica fume gave a slight increase in fluidity, but when the addition was 10%, the slump increased greatly to 235 mm. With a further increase to 15%, the increase in slump was limited and the spread decreased. Therefore, with respect to fluidity, the addition of 10% silica fume in the binder is optimum. In the past, little was known about the possible increase of fluidity in concrete mix by adding silica fume. Silica fume consists of micro ball-like grains with an average size of 0.1 μm. In a natural state, these clump together to form coarse agglomerates. During vigorous mixing, these agglomerates can be dispersed so the superplasticizer covers the surface with a layer of surface-active molecules giving rise to an electrostatic repulsive force between the particles (like the cement particles), preventing them from reflocculating. Because the ball-like silica fume particles are much smaller than the cement particles (average particle size 10 μm), they play the role of ball bearings between cement particles which improves the fluidity of the cement paste. In addition, the silica fume particles fill the pores between cement grains substituting for water, allowing it to be available to improve the fluidity of the concrete mix.

With respect to strength, 10% silica fume is also the optimum addition because with an addition of 15% silica fume, although R_{56} is higher than that with 10% addition, the increase is small.

The author has analyzed the pozzolanic effect and strength contribution of silica fume in super-high-strength concrete with the analysis procedure given in Pu Xincheng, Yan Wunan, Wang Cong et al.[25] This analysis showed that with the addition of 10%~15% silica fume, the contribution of the pozzolanic reaction to strength was 30%~34% of the 150-MPa super-high-strength concrete. Without the addition of silica fume, it is impossible to produce 150-MPa concrete with ordinary materials and conventional technology. Figure 6.8 shows the effects that the addition of 5%~15% of silica fume has on the strength of super-high-strength concrete.

6.9 THE INFLUENCE OF CURING CONDITIONS AND AGE ON THE STRENGTH OF CONCRETE

It is well known that the way in which concrete is cured has a marked effect on its microstructural formation and ultimate strength development. To

Table 6.10 Influence of silica fume addition on the fluidity and strength of super-high-strength high performance concrete

No.	$B/kg/m^3$	Binder composition/%		S/%	W/B	WRA	Fluidity/mm		28 d ρa	Compressive strength/MPa			
		C	SF			L/m^3	Slump	Spread	kg/m^3	3 d	28 d	56 d	365 d
1	650	100	0	40	0.20	28	0	—	2563	85.8	108.1	113.6	117.3
2	650	95	5	40	0.20	28	17	—	2565	89.7	126.8	129.1	131.7
3	650	90	10	40	0.20	28	235	448	2581	99.3	140.7	147.6	157.7
4	650	85	15	40	0.20	28	240	435	2569	92.3	136.9	148.0	151.0

Note: B = binder content; S = sand percentage; W/B = water binder ratio; 28 $d\rho_a$ = 28-day apparent density of the concrete.

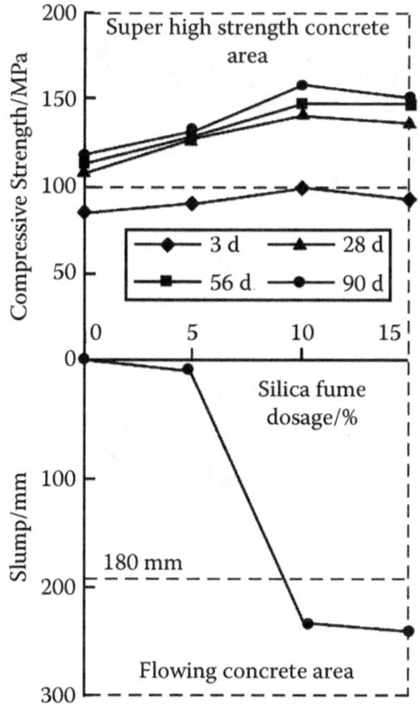

Figure 6.7 Influence of silica fume addition on the fluidity and strength of super-high-strength high performance concrete.

promote the transition of super-high-strength high performance concrete production from laboratory to building site, the author carried out a study on the influence of curing conditions on the strength development of super-high-strength high performance concrete. In this study, 525 OPC from Tenghui Cement Co, silica fume from Qinzhen, Guizhou province, crushed limestone from Chongqing Xiaoquan, medium sand, and melamine resin superplasticizer were used.

Four types of curing conditions were chosen:

1. Normal curing: after demolding, the specimens were moved to a standard curing room with temperature 20 ± 2°C and relative humidity of 90% or more.
2. Curing in water: after demolding, the specimens were moved to a water tank in a standard curing room with a water temperature of 20 ± 2°C, which is the best condition for curing.
3. Curing with moisture isolation: after demolding, the specimens were placed in a plastic bag, sealed, and cured in the laboratory workshop at varying temperatures. Such conditions imitate curing conditions for concrete encased in a steel tube.

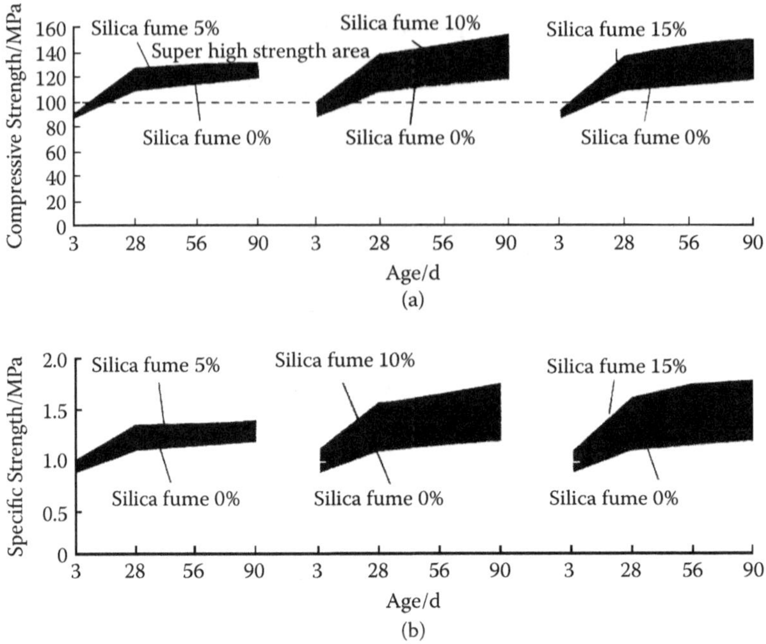

Figure 6.8 Relationship of compressive strength, specific strength, specific strength from pozzolanic effect, and contribution rate from pozzolanic effect of silica fume concrete with age: (a) relationship of compressive strength of silica fume concrete with age; (b) relationship of specific strength of concrete with age; (c) specific strength of pozzolanic effect with age; (d) contribution rate of pozzolanic effect with age.

4. Natural curing: after demolding, the specimens were kept in the laboratory workshop with free moisture and heat exchange with the natural environment: this is the most unfavorable curing condition.

Three mix proportions were chosen with water/binder ratios of 0.20, 0.25, and 0.30. The parameters of mix proportions are given in Table 6.11. For each formulation, 24 mixes with 72 specimens were made; test results are given in Table 6.12. The compressive strength of the concrete cured under normal curing conditions at different ages was taken as 100%. The relative strengths for curing conditions other than normal curing are shown as percentages of that for normal curing in Table 6.12. The compressive strength development curves of concretes with different mix proportions cured under different conditions and curing ages are shown in Figure 6.9.

The following points can be seen from Table 6.12 and Figure 6.9:

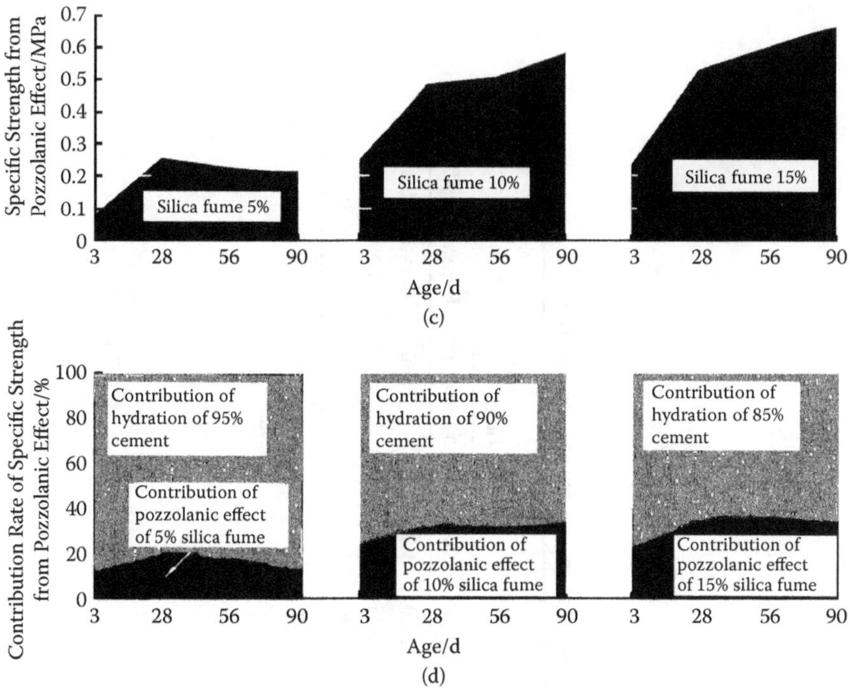

Figure 6.8 (continued)

1. No matter what curing condition was used, strength development of super-high-strength high performance concrete increased with age, even under the most unfavorable condition of natural curing. The strength grew rapidly at early ages (0~28 d) but slowed at later ages (90~365 d), which is similar to strength growth of ordinary concrete. In the experiment, there was one case where the strength at a later age was slightly lower than that at an earlier age, which may be due to the method of strength measurement, which is difficult to keep consistent, together with specimen preparation.
2. Under the conditions of water curing, there is an abundant supply of water, so the strengths of super-high-strength high performance concrete at different ages were mostly above that of concrete at the same age cured normally. This was particularly noticeable for the concrete with higher water/binder ratio (0.30), where the strength of the concrete cured in water was highest by 22.3%. When the water/binder ratio was lowered, the percentage increase of strength was reduced; for example, at a water/binder ratio of 0.25, the biggest increase was 7.6% while at water/binder ratio of 0.20, the increase was only 4.1%.

Table 6.11 Mix proportions and fluidity parameters for concrete to study curing conditions

No.	Binder content kg/m³	Composition of binder		Water content kg/m³	Sand percentage/%	W/B	WRA %	Fluidity		Forming condition
		Cement	SF					Slump	Spread	
1	650	90	10	130	40	0.20	1.9	270	660	Vibration 10 s
2	650	90	10	163	40	0.25	0.7	271	713	Tamping
3	650	90	10	195	40	0.30	0.3	80	—	Vibration 90 s

Note: SF = silica fume; W/B = water binder ratio; WRA = water-reducing admixture. Water-reducing admixture used was DSF-2.

Table 6.12 Effect of curing condition on the strength development of super-high-strength high performance concretes

				Curing age/d														
			$28d\ \rho_a$	*3*			*28*			*56*			*90*			*365*		
No.	*W/B*	*Curing*	kg/m^3	*AV/MPa*	*RV/%*	*Inc/%*	*AV/MPa*	*RV/%*	*Inc/%*	*AV/MPa*	*RV/%*	*Inc/%*	*AV/MPa*	*RV/%*	*Inc/%*	*AV/MPa*	*RV/%*	*Inc/%*
1	0.20	Nor	2545	79.6	100		126.3	100		130.6	100		148.4	100		149.2	100	
		W	2562	77.4	97.2	−2.8	130.2	103.1	+3.1	135.9	104.1	+4.1	144.5	97.4	−2.6	152.2	102.0	+2.0
		HI	2533	74.1	93.1	−6.9	131.5	104.1	+4.1	134.6	103.1	+3.1	145.1	97.8	−2.2	143.0	95.8	−4.2
		NC	2520	76.1	95.6	−4.4	116.9	92.6	−7.4	126.1	96.6	−3.4	132.9	89.6	−10.4	136.1	91.2	−8.8
2	0.25	Nor	2497	61.7	100		111.3	100		115.0	100		127.9	100		141.1	100	
		W	2509	61.3	99.4	−0.6	116.1	104.3	+4.3	123.7	107.6	+7.6	129.0	100.9	+0.9	147.9	104.8	+4.8
		HI	2409	58.0	94.0	−6.0	112.8	101.3	+1.3	112.3	97.6	−2.4	127.2	99.5	−0.5	129.7	91.9	−8.1
		NC	2477	62.6	102.1	+2.1	101.3	91.0	−9.0	111.7	97.1	−2.9	119.8	93.7	−6.3	121.2	85.9	−14.1
3	0.30	Nor	2450	51.9	100		85.2	100		91.5	100		93.3	100		95.3	100	
		W	2474	53.4	103.0	+3.0	104.2	122.3	+22.3	102.9	112.5	+12.5	104.9	112.4	+12.4	105.0	110.2	+10.2
		HI	2456	53.1	102.3	+2.3	97.1	114.0	+14.0	93.5	102.2	+2.2	96.1	103.0	+3.0	98.7	103.6	+3.6
		NC	2409	50.6	97.5	−2.5	78.6	92.3	−7.7	84.7	92.6	−7.4	85.7	91.9	−8.1	86.5	90.8	−9.2

Note: w/B = water binder ratio; 28d ρ_a = 28 d apparent density of concrete; AV = absolute value; RV = relative value; Inc = increment; Nor = normal curing; W = curing in water; HI = curing with moisture isolation; NC = natural curing.

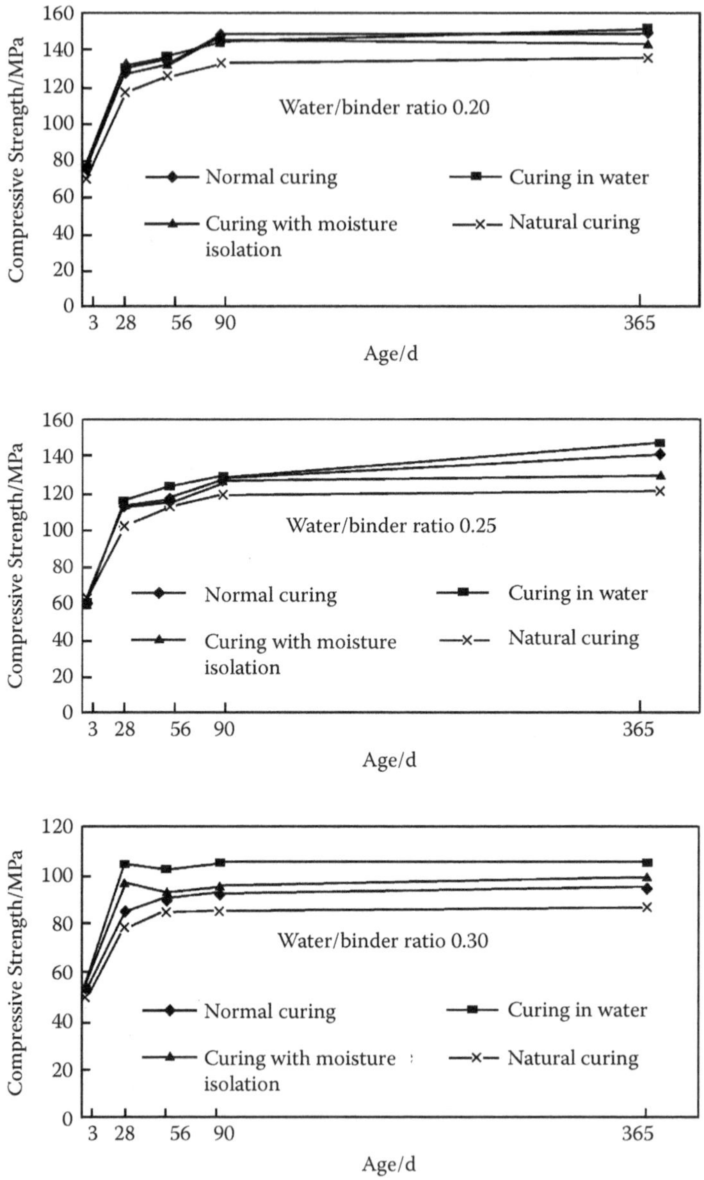

Figure 6.9 Strength development with age under different curing conditions.

This may be due to the decrease in permeability with the lowering of the water/binder ratio.

3. When the high water/binder ratio (0.30) sample was cured with moisture isolation, the increase in strength at different ages was 2.2%~14% higher than that from normal curing. In this case, although the concrete lost any addition to the water supply, the large amount of water contained in the concrete was sufficient so there was no influence on the strength development, and the strength was no lower than that cured under normal conditions. When the water/binder ratio was lowered, the strength at 28 d was higher than that under normal curing, but with increased time, its strength fell below that from normal curing at the same age; this may be due to the lack of water supply.
4. Under natural curing, the strength of concrete generally was lower than that obtained under normal curing at the same age. With increasing age, the rate at which the strength decreased increased, so that at 365 d, the strength was lowered for three water binder ratios by 9.2%, 14.1%, and 8.8%, respectively. It showed that under such curing conditions, water was lost from the concrete, which meant a lack of water for cement hydration, and consequently a reduction in concrete strength.

The natural curing condition used by the author was an extreme case, which is rarely found in construction. On the contrary, in actual construction work, moist curing is often undertaken, at least for the first few days. It can be seen from this study that even under such conditions, the reduction in strength for super-high-strength high performance concrete is limited to no more than 10% that of normal cured concrete at 28 d and 15% at 365 d. This is quite different from ordinary concrete, where according to the data in the book by A. M. Neville,[50] for ordinary concrete with a water/cement ratio of 0.5, the strength of concrete cured in air is much lower than that obtained under moisture curing (Figure 6.10), where at 28 d, the decrease is about 45% and at 90 d, 53%. It shows that the strength development of super-high-strength high performance concrete is less susceptible to moisture levels during curing than ordinary concrete. This is because the water/binder ratio of super-high-strength high performance concrete is low, giving a dense structure with few capillary pores and low permeability so that even under the natural curing given in this study, the loss of water from the concrete is reduced. Impermeability testing of super-high-strength high performance concrete showed that its impermeability grade was over S40, which illustrates the dense structure of the concrete. Although super-high-strength high performance concrete is less susceptible to moisture loss, in practice, special care should be given to moisture curing to avoid a reduction in strength, especially that due to shrinkage and cracking.

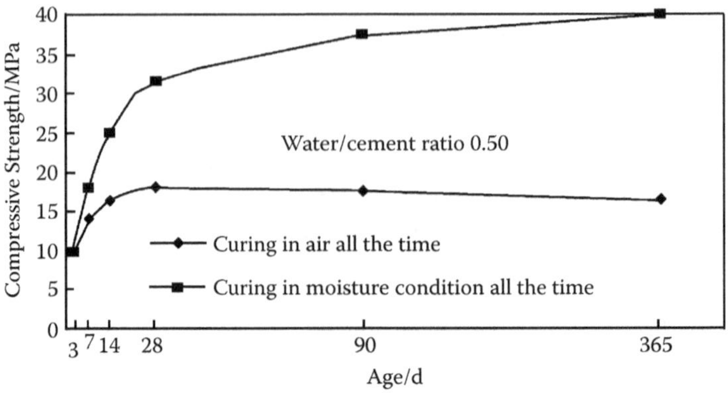

Figure 6.10 Influence of different curing conditions on the strength of ordinary concrete.

It can be concluded from the studies above that

1. Under the four conditions used in our study, the trend is for strength development in super-high-strength high performance concrete to increase with age rapidly at early ages and more slowly at later ages.
2. The super-high-strength high performance concrete cured in water develops more strength than that cured normally; the strength when cured with moisture isolation is close to that obtained with normal curing, while the strength under natural curing is lower than that under normal curing.
3. In comparison with ordinary concrete, super-high-strength high performance concrete has a dense structure, making moisture exchange with the environment difficult so it is less susceptible to moisture loss. Therefore, the range in reduction of strength for super-high-strength high performance concrete due to different curing conditions is less than that of ordinary concrete.

Chapter 7

Structure

7.1 HYDRATION DEGREE AND HYDRATION PRODUCTS OF THE BINDER

As described previously, there are four key minerals in Portland cement, namely, C_3S, βC_2S, C_3A, and C_4AF, as well as gypsum $CaSO_4{\cdot}2H_2O$. The C_3A, C_4AF, and C_3S hydrate rapidly, but βC_2S only slowly. C_3A hydrates rapidly to form C_3AH_6 but in the presence of $Ca(OH)_2$ forms $4CaO{\cdot}Al_2O_3{\cdot}19H_2O$, and if there is SiO_2 available, then a substituted hydrogarnet $3CaO{\cdot}Al_2O_3{\cdot}xSiO_2(6\text{-}2x)H_2O$ can be formed slowly in the final stage. C_3A in presence of $CaSO_4{\cdot}2H_2O$ rapidly forms tricalcium sulphoaluminate hydrate (ettringite, $3CaO{\cdot}Al_2O_3{\cdot}3CaSO_4{\cdot}32H_2O$), which coats the surface of the cement grains, retarding the setting of the cement. When the gypsum is exhausted, the tricalcium aluminate may react with ettringite to form calcium sulphoaluminate hydrate of monosulphur type ($3CaO{\cdot}Al_2O_3{\cdot}CaSO_4{\cdot}12H_2O$).

Tetracalcium aluminoferrite reacts with water to form cubic tricalcium aluminate and ferrite hydrates. If $Ca(OH)_2$ and $CaSO_4$ are present, then calcium trisulfoaluminoferrite $3CaO(Al_2O_3{\cdot}Fe_2O_3){\cdot}3CaSO_4{\cdot}32H_2O$ may be formed. When the gypsum is exhausted, calcium sulphoaluminate or ferrite hydrate of a monosulfate type may be formed.

However, C_3A and C_4AF make up no more than 25% of the mass percentage in cement and have little influence on the composition and structure of hardened cement paste. The most important minerals in cement are C_3S and C_2S, which may reach 75% mass percentage. Their hydration process will determine the composition of the hydration products and structure of the hardened cement paste. Initially, a calcium silicate hydrate known as C-S-H (I) with a lime/silica ratio of about 1.5 is formed, sometimes called CSH (B) in some literature. The final hydration product, calcium silicate hydrate C-S-H (II), has a lime/silica ratio of 1.6 to 2.0 and is also called C_2SH_2. C-S-H (I) and C-S-H (II) are poorly crystalline and close to amorphous. The hydration products of calcium silicate formed in a cement paste have small dimensions and are close to colloidal size and are sometimes called calcium silicate hydrate gel or C-S-H gel. Consequently, the main characteristics of calcium silicate hydrate gel are as follows:

1. Its composition varies continuously. The lime/silica ratio may vary from several tenths up to 2.
2. Differences in crystallization. In the series of calcium silicate hydrates, there are also some well-crystallized hydrates with a fixed lime/silica ratio, such as tobermorite ($4CaO{\cdot}5SiO_2{\cdot}5{\cdot}5H_2O$), xonotlite ($6CaO{\cdot}6SiO_2{\cdot}H_2O$), and so forth. They do not appear usually in cement paste hydrated under natural conditions but are generated by continuous hydrothermal treatment with the proper CaO/SiO_2 ratio of the raw materials.

With an active mineral admixture, the reactive SiO_2 undergoes a secondary reaction with the high Ca/Si ratio C-S-H gel and $Ca(OH)_2$ generated from hydration of cement, in the so-called pozzolanic reaction, to produce more calcium silicate hydrates and lower the Ca/Si ratio of the calcium silicate hydrates, transforming the high Ca/Si ratio C-S-H gel (CSH (II) or C_2SH_2) into a low Ca/Si ratio C-S-H gel (C-S-H (I) or CSH (B)), improving the quality and quantity of the calcium silicate hydrate binder.

Justnes[51] analyzed the degree of hydration, Ca/Si ratio of the hydration products, the length of silicon–oxygen anionic chain, the degree of the pozzolanic reaction of silica fume, the amount of free calcium hydroxide during hydration of C_3S and C_2S in a Portland cement paste (C_3S 43.5%, C_2S 32.7%, C_3A 5.9%, C_4AF 8.8%; specific surface 405 m^2/kg) with silica fume (SiO_2 94.3%, dosage 0, 8%, and 16%) using ^{29}Si nuclear magnetic resonance and thermal weight loss methods (Table 7.1). The results show

1. At all ages, by reducing the water/binder ratio from 0.4 to 0.2, the degree of hydration reduces. For example, at 28 d and a water/binder ratio 0.4, the degree of hydration reaches 66.4% (without silica fume), 52.9% (8% silica fume), 63.0% (16% silica fume); at a water/binder ratio of 0.3, the hydration degree is 45.5% (without silica fume), 37.9% (16% silica fume); at a water/binder ratio of 0.2, the hydration degree is 31.4% (without silica fume), 30.7% (8% silica fume), 27.2% (16% silica fume). The addition of silica fume lowers the degree of hydration to some extent.
2. With the addition of silica fume, the amount of free calcium hydroxide in cement paste is reduced compared to that without silica fume; with an increase in the addition, more $Ca(OH)_2$ is used which increases with age; with an increase in water/binder ratio, the amount of free calcium hydroxide in cement paste is increased, both for the pastes with and without silica fume. At a water/binder ratio of 0.2, in the paste with 16% silica fume, the mass percentage of free calcium

Table 7.1 Index of hydration character of cement pastes with or without silica fume

Mass percentage of ingredients in cement paste (%)				Index of hydration character					
Cement	Silica fume	Water/binder ratio W/(C+S)	Age (d)	Hydration degree of $C_3S + C_2S$/%	Average length of Si-O anion chain, number of SiO_4^{4-}	CH content (%)	Degree of pozzolanic reaction for silica fume (%)	C/S for calcium silicate hydrate	Mass ratio of water to reacted CSH $\frac{W}{C+S}$
100	0	0.4	1	39.8	2.30	10.45		1.11	0.28
100	0	0.4	3	50.7	2.58	12.26		1.27	0.27
100	0	0.4	28	66.4	2.71	13.49		1.32	0.24
100	0	0.4	442	83.1	3.16	15.65		1.36	0.36
92	8	0.4	1	40.5	3.08	8.45	40.6	0.83	0.22
92	8	0.4	3	46.4	3.08	9.65	43.1	0.87	0.24
92	8	0.4	28	52.9	3.00	6.83	100	0.85	0.23
92	8	0.4	126	64.7	3.41	5.95	100	0.97	0.34
92	8	0.4	442	63.1	3.16	6.27	100	0.95	0.24
84	16	0.4	1	33.4	2.82	7.12	13.7	0.77	0.29
84	16	0.4	3	44.8	3.01	6.94	44.8	0.67	0.21
84	16	0.4	28	63.0	4.02	2.56	94.5	0.69	0.20
84	16	0.4	126	61.3	4.02	0.75	100.0	0.74	0.24
100	0	0.3	28	45.5	2.94	9.92		1.44	0.25

continued

Table 7.1 (continued) Index of hydration character of cement pastes with or without silica fume

Mass percentage of ingredients in cement paste (%)				*Index of hydration character*					
Cement	*Silica fume*	*Water/binder ratio W/(C+S)*	*Age (d)*	*Hydration degree of $C_3S + C_2S$/%*	*Average length of Si-O anion chain, number of SiO_4^{4-}*	*CH content (%)*	*Degree of pozzolanic reaction for silica fume (%)*	*C/S for calcium silicate hydrate*	*Mass ratio of water to reacted CSH $\frac{W}{C+S}$*
84	16	0.3	28	37.9	3.61	2.10	74.4	0.60	0.21
100	0	0.2	1	12.1	2.00	5.01		0.011	0.401
100	0	0.2	3	23.1	2.57	5.24		1.28	0.26
100	0	0.2	28	31.4	3.10	6.23		1.53	0.18
100	0	0.2	442	30.6	3.19	5.76		1.73	0.13
92	8	0.2	28	30.7	3.61	1.71	74.6	0.94	0.24
92	8	0.2	442	33.2	3.69	1.94	90.1	0.87	0.35
84	16	0.2	1	13.2	2.83	3.12	13.4	0.44	0.28
84	16	0.2	3	13.9	2.74	3.23	22.4	0.28	0.34
84	16	0.2	28	27.2	3.86	0.63	66.8	0.54	0.22
84	16	0.2	126	25.3	4.00	0.00	79.0	0.44	0.32
84	16	0.2	442	23.6	4.24	0.00	83.2	0.42	0.39

Note: At water/binder ratio of 0.20, superplasticizer was added.

hydroxide in cement paste at age 28, 126, and 442 d are 0.63%, 0, and 0, respectively.

3. As the paste ages, the average length of the silicon oxygen anionic chain increases, for example, at a water/binder ratio of 0.2, and age 1, 3, 28, and 442 d, there are 2.0, 2.57, 3.1, and 3.19 siloxy tetrahedrons, respectively, per chain.
4. With an increase in age, the Ca/Si ratio of the calcium silicate hydrate in a hardened cement paste without silica fume increases; at a water/cement ratio of 0.2, the Ca/Si ratios at 28 and 442 d are 1.53 and 1.73, respectively; for the hardened cement paste with 8% of silica fume, the Ca/Si ratios at related ages is 0.94 and 0.87, respectively; with 16% silica fume, the Ca/Si ratios at the same ages are 0.54 and 0.42, respectively. With an increase in water/binder ratio, the Ca/Si ratio increases.
5. With increasing age, the degree of pozzolanic reaction of the silica fume (% of the silica fume reaction) increases; for example, at a water/binder ratio of 0.2 and with an addition of 16% silica fume, at the age of 1, 3, 28, 126, and 442 d, the degrees of reaction are 13.4%, 22.4%, 65.8%, 79.0%, and 83.2%, respectively; with an increase in water/binder ratio, the degree of reaction of silica fume increases, for example, at a water/binder ratio of 0.4, the degrees of reaction are 13.7%, 44.8%, 94.5%, and 100%, respectively.

As mentioned above, the addition of an efficient active mineral admixture will generally lower the degree of hydration of the cement. The lower the water/binder ratio, the obvious the reduction. This is because the pozzolanic reaction of active mineral admixture gives rise to a large amount of C-S-H gel. The presence of adsorbed and interlayer water in the gel reduces the amount of free water available for the hydration reaction leading to a reduction in the degree of hydration of the cement. However, the quantity of gel formed by hydration per 1 g of cement increases greatly, the quantity of non-evaporable water in the whole system is enhanced.

The study by Sellevold and Justnes[52] showed that the hydration degree of the binder, hs, can be calculated from the ratio of the quantity of non-evaporable water, W_{ns}, taking part in the hydration, to the quantity of binder (cement plus silica fume) in the system reduced by 0.25 (it drops slightly when the silica fume content increases), so

$$hs = \frac{W_{ns}}{(C+S)0.25} \tag{7.1}$$

where

C = mass of cement
S = mass of silica fume

Table 7.2 Hydration degree of paste with water/binder ratio of 0.20 with different additives at 28 d

Type of binder	*Water content of CH per 1 g of cement (mg)*	*Water content of H per 1 g of cement (mg)*	*Non-evaporable water content per 1 g of cement (mg)*	*Hydration degree hs (%)*
C = 100%	26.67	88.33	115	46
C = 80%, YX120 = 20%	18.01	129.90	147.91	59.2
C = 80%, YX600 = 20%	14.99	138.18	153.17	61.3
C = 80%, PX120 = 20%	12.55	147.22	159.77	63.9
C = 80%, SL300 = 20%	19.48	116.22	134.62	53.8
C = 80%, FA300 = 20%	18.40	122.68	141.8	56.4
C = 80%, SF = 20%	12.80	141.98	154.78	61.9

Note: C-525 OPC from Chongqing Cement Plant, YX120, YX600, PX120 are three kinds of X powder with different surface area, SL300 = slag; FA300 = fly ash; SF = silica fume; the chemical composition is given in Tables 6.1 and 6.2.

If W_{ns} is measured, the hydration degree of the system can be calculated. Wan Chao-jun[53] has measured water content of CH (the bound water of $Ca(OH)_2$) and bound water of the gel H for samples with the different active minerals at 28 d. According to Equation (7.1), W_{ns} in the sample is made up of CH plus H so the hydration degree of the sample can be determined, as shown in Table 7.2. It can be seen that by adding an active mineral admixture, the content of bound water of the gel in the system increased significantly (31.6% to 66.7%), and with the increase in the content of gel, the degree of hydration of the system increased by 17% to 38.9%.

Wang Yongwei[54] has studied the hydration products of hardened super-high-strength cement paste prepared from Chongqing 525 cement, with the addition of Qinzheng silica fume, slag from the Chongqing Steel Company, fly ash from the Chongqing Power Plant, or X powder and UEA (chemical compositions of these are given in Tables 6.1 and 6.2) by X-ray diffraction and scanning electronic microscopy. The mix proportion of the cement pastes, their fluidity, and strength are given in Table 7.3.

The X-ray diffraction study (Figure 7.1) shows

1. In the hardened cement paste with a water/binder ratio of 0.20, a large amount of unhydrated cement clinker grains remained. It can be seen from the X-ray diffraction traces in Figure 7.3, the high peaks at 0.304 nm, 0.278 nm, 0.274 nm, 0.270 nm, 0.265 nm, 0.261 nm, 0.232 nm, 0.218 nm, 0.177 nm, 0.163 nm, and 0.149 nm arise from the cement. In the six traces shown in Figure 7.1, the characteristics of the peak (d = 0.260 to 0.304 nm) around 2θ = 30° is remarkable. The characteristic peaks of C_3S (0.279 nm, 0.265 nm, 0.307 nm, 0.284 nm, 0.179 nm or 0.276 nm, 0.274 nm, 0.259 nm, 0.302 nm,

Table 7.3 The mix proportions and test results of fluidity and strength of cement paste

	Mass percentage of components in paste (%)								*Fluidity (mm)*		*Strength at 56d*
No.	*Cement*	*SF*	*SL*	*FA*	*X Powder*	*UEA-H*	*W/B*	*WRA (%)*	*Slump*	*Spread*	*(MPa)*
W1	100	0	0	0	0	0	0.20	2.0	48		110.4
W2	70	10	20	0	0	0	0.20	2.0	290	1050	135.9
W3	70	10	0	20	0	0	0.20	2.0	290	1100	132.9
W4	70	10	0	0	20	0	0.20	2.0	290	950	133.4
W5	78	10	0	0	0	12	0.20	2.0	270	620	127.7
W6	90	10	0	0	0	0	0.20	2.0	290	950	128.9
W7	90	10	0	0	0	0	0.25	1.68	290	1150	116.9
W8	90	10	0	0	0	0	0.30	0.7	290	1350	93.3

Note: SF = silica fume; SL = slag; FA = fly ash. Strength at 56 d = the compressive strength of 100 mm × 100 mm × 100 mm cube at 56-day age. The slump is the same as for the concrete but without tamping in layers. The spread is the average of two perpendicular diameters of the disk formed after raising the slump cone.

and 0.218 nm) and β- C_2S (0.280 nm, 0.274 nm, 0.278 nm, 0.261 nm, 0.273 nm, 0.272 nm, and 0.198 nm) are apparent. While there are also characteristic peaks for C_3A (0.270 nm, 0.191 nm, 0.156 nm, 0.408 nm, 0.274 nm, 0.220 nm, 0.424 nm, and 0.135 nm) and C_4AF (0.263 nm, 0.277 nm, 0.192 nm, 0.267 nm, 0.204 nm, 0.724 nm, 0.181 nm, and 0.157 nm), these are not apparent, which may be due to the low content of C_3A and C_4AF in the cement as well as their high hydration rates.

Using the height of the peak at 0.278 nm, the order is W5 > W6 > W4 > W2 > W1 > W3; using 0.274 nm gives W5 > W4 > W2 > W6 > W3 > W1, while at 0.261 nm the order is W5 > W2 > W6 > W4 > W3 > W1. These three peaks are due to β-C_2S and C_3S.

The height of this peak is determined jointly by the degree of hydration of the cement and the cement content, so it may be accepted that for the degree of hydration, the order is W1 > W3 > W2 ≈ W4 ≈ W6 > W5. This conclusion is also consistent with the influence of the additive on the degree of hydration of the cement. The decrease in the degree of cement hydration is mainly due to the high water requirement of UEA-H and the large amount of water needed in the formation of ettringite.

2. In the six X-ray diffraction patterns in Figure 7.1, there is a peak around $2\theta = 30°$ (d = 0.25 to 0.35 nm) region. Taylor takes the view[55] that a diffuse peak in the region of 0.27 to 0.31 nm and a peak at 0.182 nm are characteristic of the pattern of C-S-H. The characteristic peaks determined by Taylor for amorphous CSH (I) ((0.8 to 1.5) $CaO.SiO_2 \cdot (0.5\text{–}2.0)H_2O$) were 1.25 nm, 0.53 nm, 0.304 nm, 0.280 nm, 0.24 nm, 0.21 nm, 0.182 nm, 0.167 nm, 0.152 nm,

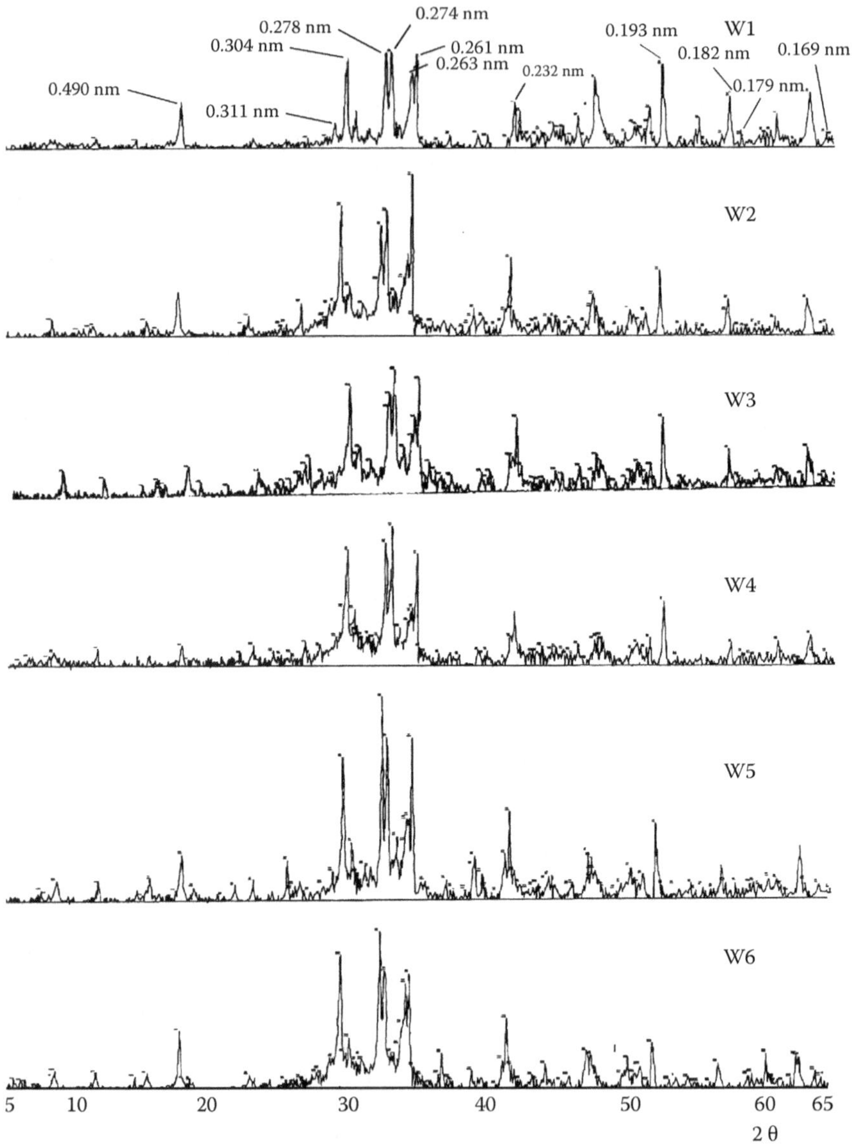

Figure 7.1 XRD diagram of the paste with different additives. (From Wang Yongwei, "Study on Composition, Structure and Shrinkage and Its Compensation of Super-High-Strength High Performance Concretes," master's dissertation, Chongqing: Chongqing University, 2001 [in Chinese].)

0.140 nm, 0.123 nm, 0.117 nm, 0.111 nm, and 0.107 nm, with the strongest peaks at 1.25 nm, 0.304 nm, 0.280 nm, and 0.182 nm, while those determined by Seikin[56] were 0.305 nm, 0.262 nm, and 0.183 nm. In the patterns in Figure 7.1, peaks at 0.304 nm, 0.280 nm, and 0.182 nm are apparent, which illustrates that there is a significant amount of C-S-H (I) gel in the hardened cement paste.

It can be seen from Figure 7.1 that with the addition of the additives W2, W3, W4, W5, and W6, the peak is larger than that in the pure cement paste (W1), especially when silica fume and slag (W2), silica fume and fly ash (W3), and silica fume and X powder (W4) are combined. This is due mainly to the increase of the amorphous C-S-H (I) gel formed after addition of mineral admixtures.

3. In all six X-ray diffraction patterns, there are CH peaks at 0.263 nm, 0.490 nm, 0.193 nm, 0.179 nm, 0.311 nm, and 0.169 nm. Using the peak at 0.490 nm, the CH content can be seen to be highest in the pure cement paste (W1), while it is reduced in the pastes with combined additions of fume and slag (W2), silica fume and fly ash (W3), and silica fume and X powder (W4), especially for W3 and W4. There, the reduction is significant as X powder and fly ash have little CaO. Single additions of silica fume (W6) and a combined addition of silica fume plus expansive agent (W5) make almost no difference. The peak at 0.190 nm is even a little higher than for W1, but examining the peaks at 0.311 nm and 0.193 nm suggests that it is lower in W6 than W1, but this may be due to the superimposition of specific peaks.
4. It can be seen from Figure 7.2 that with an increase of water/binder ratio, the content of unhydrated particles of cement clinker is reduced. In the cement paste with water/binder ratio 0.30 (W8), the content of clinker drops rapidly. This is closely related to the degree of hydration as described above. In the three X-ray diffraction patterns shown in Figure 7.2, the diffuse peaks around the $2\theta = 30°$(d = 0.25 to 0.35 nm) reduce as the water/binder ratio is increased. The characteristic peaks at 0.304 nm and 0.182 nm reduce with an increase in the water/binder ratio, illustrating how the amorphous C-S-H (I) gel content diminished with an increase in the water/binder ratio. With an increase in water/binder ratio, there was no obvious change in the characteristic peaks of CH so no obvious change in CH content. In three of the mix formulations, the initial lime/silica mole ratios (C/S) are the same; the degree of hydration of the system increased with an increase in the water/binder ratio, illustrating that the C/S ratio of the C-S-H gel in cement paste increases with an increase in water/binder ratio. As the water/binder ratio was raised, the calcium silicate hydrate gel transformed gradually from a low alkali type to a high alkali one, reducing the amount of amorphous C-S-H (I) gel with an increase of water/binder ratio.

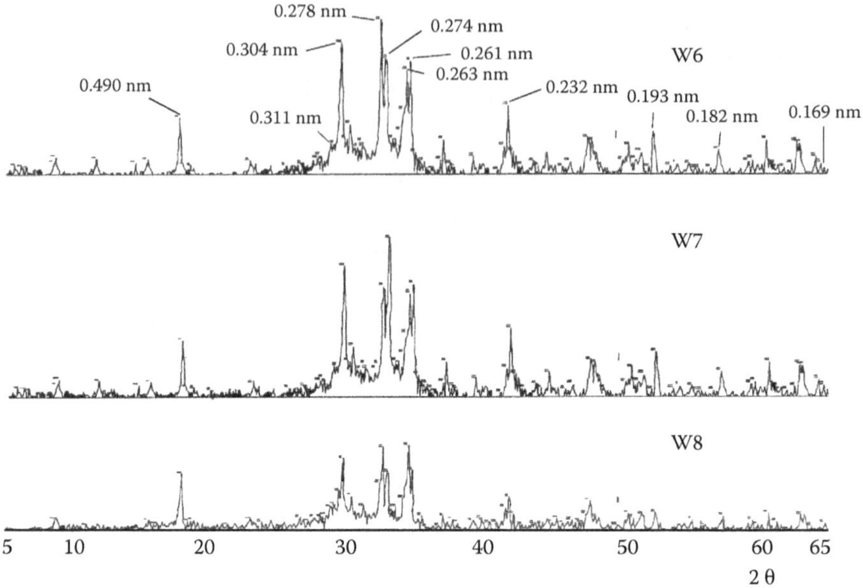

Figure 7.2 Patterns for pastes with different water/binder ratio.

Figure 7.3 XRD pattern of the cement.

The scanning electron microscope study in Figure 7.4 showed there was a significant difference in the microstructure of the hardened cement pastes containing different additives. In the pure cement paste (W1), there was a lot of fibrous material with a loose structure, and little overlap of the fibers. In the hardened paste with 10% silica fume + 20% slag (W2), there was fibrous material in many places, and in comparison with W1, the fibers were much thinner and distributed evenly, overlapping into a dense network. Where there was no fibrous material, the gel was dense and folded where there was no fiber. The fiber-like substances of W2 were similar to

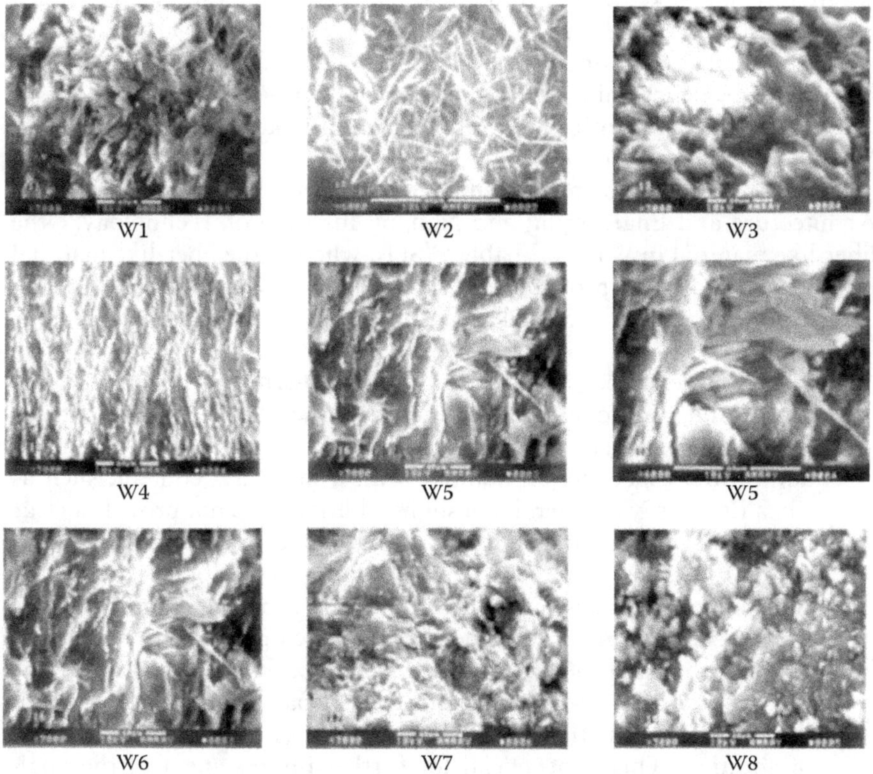

Fig 7.4 SEM micrographs of hardened super-high-strength cement paste.

those of C-S-H (II) given in the literature[19] but were larger in dimensions than those given by Chongqing Institute of Architecture and Engineering and Nanjing Institute of Technology.[19] In the hardened paste with 10% silica fume + 20% X powder (W4), the cement paste was in a flocculated form that was dense with overlaps between the flocculated clumps.

In the hardened paste with 10% silica fume + 12% UEA-H (W5), ettringite crystals coagulated in the pores and openings of the paste. The hardened paste with only 10% silica fume (W6) had a denser structure than that of W1, but was looser than that containing an addition of other additives (W2, W3, and W4). In the hardened paste of W6, there were a lot of flocks, which may consist of fine elongations interwoven and distributed in different places.

With different water/binder ratios, the microstructure of the cement paste was different. The hardened paste containing 10% silica fume and a water/binder ratio of 0.20 (W6) gave a dense gel with a surface that was covered by flock-like needles. In the hardened paste containing 10% silica fume and a water/binder ratio of 0.25 (W7), the cement gel was denser, with short flocks containing rare ball-like material. The structure of the hardened paste containing 10% silica fume and a water/binder ratio 0.30 (W8) is much looser, with the gel in the form of flakes and balls similar to W7 but in greater quantities.

Based on the results mentioned above and in Chongqing Institute of Architecture and Engineering and Nanjing Institute of Technology,[19] the fiber-like material in W1 is probably C_2SH_2, whereas the fiber-like material in W2, W4, and W6 is probably C-S-H (I).

Based on the results discussed above the following can be concluded:

1. In the hardened cement paste of super-high-strength concrete, a low water/binder ratio cannot effectively reduce the $Ca(OH)_2$ content in the concrete. The most effective way is by addition of a mineral admixture, especially an additive with a low CaO content, such as silica fume or X powder, but a slight addition does not provide a clear outcome. The combined addition of 10% silica fume + 20% X powder or 10% silica fume + 20% fly ash will provide good results.
2. Addition of a fine active mineral admixture is beneficial in increasing the C-S-H gel content in cement paste which gives rise to increased polymerization of the C-S-H gel and a reduction in the C/S ratio. The C/S ratio of a super-high-strength cement paste containing an additive is usually less than 1.5 and decreases with increasing amounts of addition. This is beneficial for further improvement of the quality of calcium silicate hydrate, enhancing the strength of super-high-strength cement paste, and improving its stability.
3. There are many unhydrated cement particles in the hardened paste of super-high-strength concrete. The arrangement of these particles within the hydration products and particles of coarse and fine aggregates has an important effect on the concrete's strength and durability. The addition of mineral admixtures of various sizes in the proper quantity for the purpose intended is beneficial to the further compaction of the concrete.
4. The addition of the different additives to concrete gives rise to a different microstructure within the hydration products of hardened cement paste with a denser inner structure than that of the pure cement paste.
5. By integrating the characteristics of the composition of hydration products from different binder systems the quality of calcium silicate hydrate can be improved and the CH content in the concrete lowered and the strength and durability enhanced. Superfine additives with a high content of active SiO_2, such as silica fume or X powder, should be added to concrete to lower the lime/silica mole ratio in the mix proportions of components as well as lowering the water/binder ratio as

much as possible. As well, the combined addition of different additives can be effective; for example, the combination of silica fume with slag, X powder, or fly ash can enhance the strength to a great extent.

7.2 PORE STRUCTURE OF CEMENT PASTE IN SUPER-HIGH-STRENGTH HIGH PERFORMANCE CONCRETE

Pores are an important part of the microstructure of concrete as the total porosity and distribution of the various pore sizes in concrete have a significant influence on the concrete macrostructure; and of these, the distribution of various pore sizes has more impact. Wu Zhongwei[10] presented a concept relating the classification of pores and the partial porosity, e_i, and its influencing factor, Xi, as shown in Figure 7.5. Pores with diameter 2.5 to 20 nm were taken as non-harmful, with 20 to 50 nm less harmful. By increasing the pores with a diameter less than 50 nm and reducing the pores over 50 nm, and in particular over 100 nm, the performance of the concrete would be improved.

Metha[10] found that only the pores with a diameter of more than 135 nm would affect the strength and permeability of the concrete. Teramura and Sakai[10] take the view that the pores that have the greatest influence on the strength and permeability of the concrete are medium capillary pores with diameter 10 to 50 nm and coarse capillary pores with a diameter 50 to 10,000 nm.

In all mix formulations of super-high-strength high performance concrete, a superplasticizer is added to ensure good fluidity of the concrete,

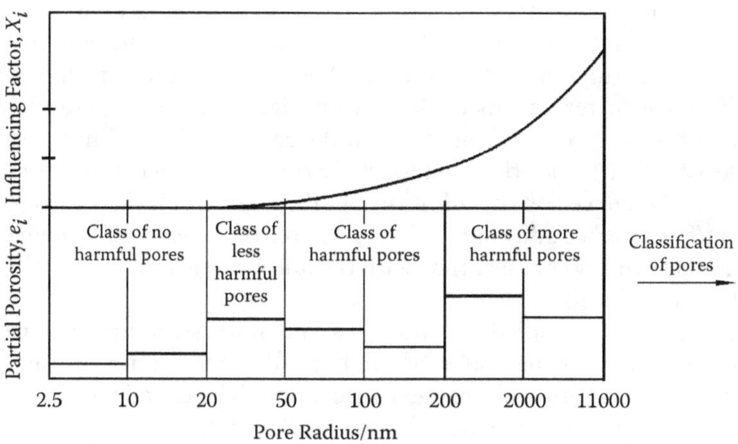

Figure 7.5 Relationship between pore class, partial porosity, and influencing factor.

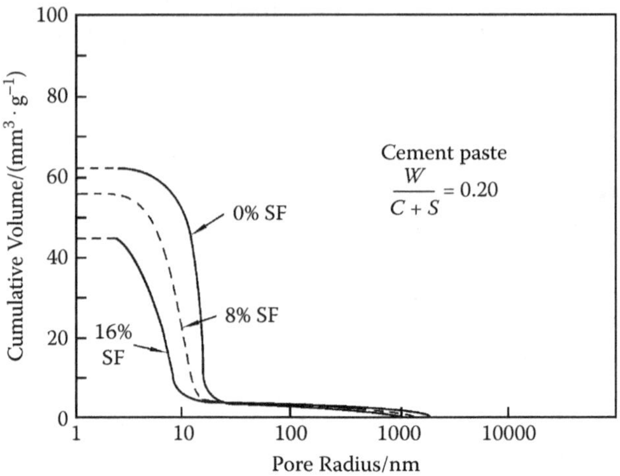

Figure 7.6 Pore size distribution of cement paste determined by a mercury porosimeter. (From E. J. Sellevold, H. Justnes et al., High Strength Concrete Binders Part B: Non-Evaporable Water, Self-Desiccation and Porosity of Cement Pastes with and without Condensed Silica Fume, Fly Ash, Silica Fume, Slag and Natural Pozzolans in Concrete, in Proceedings of the Fourth International Conference, Istanbul, 1992, Vol. II.)

allowing a low water/binder ratio (0.3 to 0.2) mix that can be compacted with existing compaction and forming methods, giving a total porosity that is lower than that of ordinary and high strength concretes. In addition, in the formulations for super-high-strength concrete, superfine active mineral admixtures should be added, their filling effect lowering the total porosity of the concrete and making the pores finer.

Figure 7.6 illustrates the integral of the pore distribution curve of hardened cement pastes with a water/binder ratio of 0.2 and addition of 0, 8%, and 16% silica fume as determined by Sellevold, Justnes et al.[52] using a mercury porosimeter. It can be seen from Figure 7.6 that there were few pores >30 nm with most of the pores in the range of 3 to 20 nm. With the addition of silica fume, the pore size distribution moved toward smaller pore sizes; the curve for the addition of silica fume at 16% moving more than that for an 8% addition making the pores of super-high-strength concrete finer. It can also be seen that with the addition of silica fume, the total porosity was reduced.

Wang Yongwei[57] studied the porosity and pore structure of hardened cement paste in super-high-strength high performance concrete using mercury intrusion porosity and nitrogen adsorption. The chemical composition and surface area of the raw materials used for the study are given in Tables 6.1 and 6.2. Both cement paste and concrete specimens were used and the mix proportions, fluidity, and strength of the cement paste specimens are given in

Table 7.4 Mix proportions, fluidity, and strength of super-high-strength high performance concrete

No.	Binder composition (%)						W/B	WRA (%)	Fluidity (mm)		Strength at 56 d (MPa)
	Cement	SF	SL	FA	X Powder	UEA-H			Slump	Spread	
W11	100	0	0	0	0	0	0.20	2.5	0	—	97.5
W12	70	10	20	0	0	0	0.20	2.0	90	—	146.5
W13	70	10	0	20	0	0	0.20	2.0	150	—	134.7
W14	70	10	0	0	20	0	0.20	2.0	50	—	136.1
W15	78	10	0	0	0	12	0.20	2.0	0	—	128.5
W16	90	10	0	0	0	0	0.20	2.0	20	—	130.7
W17	90	10	0	0	0	0	0.25	1.68	250	680	128.3
W18	90	10	0	0	0	0	0.30	0.7	245	610	114.9

Note: SF = silica fume; SL = slag; FA = fly ash. Strength at 56 d = the compressive strength of 100 mm × 100 mm × 100 mm cube at 56-d age. The spread is the average of two perpendicular diameters of the disk formed after raising the slump cone. W11–W18 concretes prepared with pastes of W1–W8 in Table 7.3.

Table 7.3 and for the concrete in Table 7.4. The specimens were prepared as 100 mm × 100 mm × 100 mm cubes and cured under standard curing conditions for 56 d. After testing the strength, fine particles under 5 mm were obtained by sieving, and immersed in anhydrous alcohol for 1 d to stop the hydration, then these particles were taken out and dried at 60°C to constant weight for measurement. Measurement by mercury penetration method was carried out on a Micromeritics Auto-pore 9200 type pressure porosimeter, and measurement by nitrogen adsorption method was on a ZXF-2 adsorption meter. The results of the measurements are shown in Table 7.4.

7.2.1 Mercury intrusion porosity (MIP)

Test results by MIP are given in Figures 7.7 and 7.8. It can be seen that the total pore volume measured for W6 was 0.0625 ml/g, for W7 0.0586 ml/g, and for W8 0.0704 ml/g. When the pore radius, R, was greater than 20 μm, the volume of mercury penetrated for samples W7 and W8 was zero, that is, there were no pores with radius more than 20 μm in the cement paste of W7 and W8; while there was a significant volume of pores with R > 20 μm (about 0.010 3 ml/g) in W6. These pores belonged to air pores and may be due to high viscosity of the cement paste with the low water/binder ratio of 0.20, where it is difficult to remove air. The pore volume for the pores in the range 0.0189 to 20 μm in W6, W7, and W8 was approximately the same (in W6 ~0.0157 ml/g, in W7 ~0.0161 ml/g, and in W8 ~0.0173 ml/g). With the increase of water/binder ratio, the pore volume increased slightly with the pore size distribution remaining about the same. For pores below 0.0189 μm, decreasing the water/binder ratio reduced the pore volume

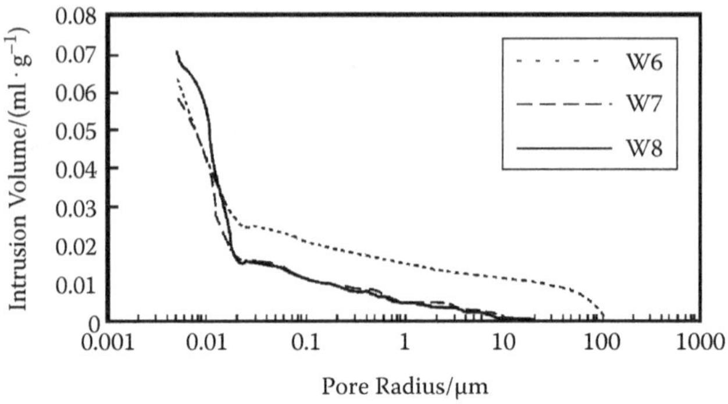

Figure 7.7 Relationship between volume of mercury penetration and pore size.

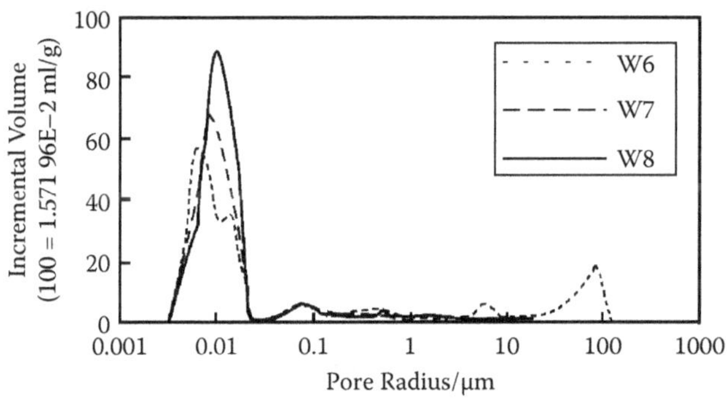

Figure 7.8 Pore size distribution of hardened super-high-strength high performance cement paste.

remarkably (in W6 to about 0.0365 ml/g, in W7 about 0.0424 ml/g, and in W8 about 0.0531 ml/g). This is related to the degree of hydration and the dense structure of the hydration products formed due to the reduction of the space normally due to *free water* in cement paste with reduction of water/binder ratio.

The lower the water/binder ratio and degree of hydration, the smaller the amount of C-S-H gel that will form leading to a reduction in pore volume. It can also be seen that for pores with $R < 0.0189$ μm, reducing the water/binder ratio changes the pore size distribution reducing the pore size of the most probable pores. For W6 the pore diameter of the most probable pores was 0.00652 μm (6.52 nm), for W7 was 0.00833 μm (8.33 nm), and W8 was 0.0106 μm (10.6 nm). This illustrates that with a reduction in water/binder ratio, the pore size became finer.

Generally speaking, the pore structure of a super-high-strength high performance cement paste is greatly improved compared to that of ordinary concrete. Most of the pores are fine. In W6 the volume of pores with R < 0.0189 μm is about 0.0365 ml/g making up 58.4% of the total volume, while if the pores more than 20 μm are excluded it makes up about 69.9% of the total volume. For W7, pores below 0.0189 μm occupy 0.0424 ml/g making up 72.5% of the total volume, and for W8 it is about 0.053 ml/g and makes up 75.4% of the total volume. According to Powers,[58] assuming the specific volume of hydrated cement is 0.414, then the porosity of W6 is 15.1%, and if the pores more than 20 μm are excluded, it makes about 12.6%; for W7, the porosity is 14.1% and for W8, 17%. The porosity of a hardened cement paste with a water/binder ratio of 0.09 formed at normal temperature and a pressure of 300 MPa measured by Xu Zhongzi, Shen Yang et al.[59] is 12.38%. If the air can be removed effectively during formation so as to exclude the air pores of more than 20 μm, cement paste with a water/binder ratio of 0.2 formed at normal temperature and pressure can have the same density.

7.2.2 Test by nitrogen adsorption

The pore volume measured by mercury intrusion shows that most of the pores of super-high-strength cement paste are concentrated below a radius of 18.9 nm. Therefore, nitrogen adsorption with its effective range of 0.5 to 30 nm was used to investigate the pore structure of hardened cement paste with different additives. The results measured by nitrogen adsorption show that for W1, the peak diameter of the most probable pore is 2.079 nm, for W2 is 2.063 nm, for W3 is 2.057 nm, for W4 is 2.032 nm, for W5 is 2.258 nm, for W6 is 2.035 nm, for W7 is 2.081 nm, for W8 is 2.038 nm, for W16 is 2.015 nm, and for W18 is 2.057 nm. Consequently, pores around 2 nm, where the most probable peaks appear, are not related to the water/binder ratio or dosage of active mineral admixtures. The appearance of such a peak may be related to the cement gel, where the most probable value of internal pores of cement gel measured by nitrogen adsorption is approximately 2 nm, which is close to 1.8 nm, the average hydraulic radius of gel internal pores as presented by Powers.[10] The average radius of the 0- to 30-nm pores for W1 is 5.122 nm, for W2 is 4.633 nm, for W3 is 4.238 nm, for W4 is 6.038 nm, for W5 is 3.526 nm, for W6 is 3.653 nm, for W7 is 4.498 nm, for W8 is 4.492 nm, for W16 is 4.091 nm, and for W18 is 6.164 nm. Except for W4, the average pore radius of a hardened cement paste containing additives is smaller than that of a pure cement paste. The most significant are those containing silica fume alone or silica fume + UEA-H. For the combined addition of silica fume + slag, silica fume + fly ash, and silica fume + X powder, the average pore radius of the hardened cement paste is larger than that of the paste with silica fume alone, which may be due to the smaller particle size of silica fume (about 0.1 μm). When the concrete sample is compared with that of cement paste (W16 with W6

Table 7.5 Volume of pores of super-high-strength high performance concrete and cement paste for different pore size intervals in 0.1-nm (Å) units: × 10^{-3}ml/g

No.	*5–20*	*20–30*	*30–40*	*40–50*	*50–100*	*100–150*	*150–200*	*200–250*	*250–300*
W1	1.197	2.985	1.406	0.921	6.692	2.595	3.154	1.044	0
W2	0	5.509	2.213	1.047	6.265	2.792	1.577	0	0.589
W3	0.260	8.690	4.150	2.908	7.156	3.048	2.646	0	1.154
W4	0.293	2.158	1.193	0.977	10.56	3.013	1.781	0	0
W5	0	17.24	7.952	2.220	6.832	1.407	2.347	0	1.985
W6	2.303	8.971	2.489	1.902	8.431	3.912	1.867	0	0.134
W7	0.690	7.943	1.360	4.119	9.813	3.257	1.268	0	1.543
W8	0	1.698	6.546	3.098	13.24	13.08	3.625	3.420	0
W16	0.449	6.112	2.077	1.668	4.174	3.600	0	0.923	0
W18	0	3.029	0	2.380	7.726	1.674	2.783	0	2.404

and W18 with W8), the average pore radius is increased particularly at a water/binder ratio of 0.3. This may be related to the rather open structure of the interface between the cement paste and aggregate. The pore volumes in the different pore size ranges for different mix proportions measured by nitrogen adsorption are presented in Table 7.5.

It can be seen from Table 7.5 that differences in the gel contents produced during hydration of the different mixes gave rise to differences in the total pore volumes measured by nitrogen adsorption. By comparing W1 with W6, which contains silica fume, the volume of gel pores under 3 nm increased greatly, while the capillary pores over 15 nm diminished with the pores over 20 nm disappearing. It can be seen that in samples with a combined addition of silica fume + slag (W2) and silica fume + fly ash (W3), the volume of gel pores less than 3 nm is reduced in comparison with that with silica fume alone (W6), but is more than that for the pure cement paste W1. This may be due to the low activity of ground slag and ground fly ash resulting in a low degree of reaction where the volume of water consumed is small, so part of the slag and/or fly ash only plays the role of filler. The real water/binder ratio of the hardened cement paste is higher than the initial water/binder ratio at mixing, so the volume of gel pores in W2 and W3 are smaller than those for W6. Meanwhile, due to the pozzolanic effect of the mineral admixtures in W2 and W3, the volume of gel increases and the volume of gel pores in W2 and W3 is higher than in W1. The volume of gel pores in W4 is much less than in a paste with addition of silica fume alone or in neat cement paste, but there is an increase of pores in the 5- to 10-nm range in W4, while the volume of pores greater than 10 nm is less than in neat cement paste and may be related to the addition of X powder. It can be seen from W5 that with the addition of the expansive admixture UEA-H, the content of gel pores increased, and that the volume of capillary pores decreased. This may be due to (1) a large amount of ettringite formed during the expansive reaction filled the capillary pores and decreased the capillary pores; and (2) there were large amounts of active SiO_2 and Al_2O_3 in the expansion admixture UEA-H, which underwent a secondary reaction with the Ca(OH) formed during hydration of cement to form more gel leading to an increase in gel pores. It can be concluded that with an increase in the water/binder ratio, the volume of large gel pores of more than 3 nm in W8 increased, compared to W7 and W6, while the volume of pores smaller than 3 nm in W8 was less than that of W6. The volume of pores less than 2 nm increased in W6.

7.3 INTERFACE STRUCTURE OF SUPER-HIGH-STRENGTH HIGH PERFORMANCE CONCRETE

Concrete is an artificial stone made from binding the aggregate with a cement paste into a monolithic entity. There are large binding interfaces between the cement paste and the aggregate—the structure and properties of which determine to a great extent the properties of concrete as a whole.

During the forming process of ordinary concrete, due to the high water/cement ratio, the adsorption of hydrophilic aggregate to the water is more than the cohesion of the water inside the cement paste, so that a film of adsorbed water is formed on the surface of the aggregate. This leads to a higher water/binder ratio in the paste near the aggregate surface than in the paste itself, with the result that the hydration products formed are different to those in the paste itself, and this zone is called the *interface transition zone* (Figure 7.9). In the transition zone, usually ettringite and coarse plate-like crystals of oriented calcium hydroxide are formed within a structure that is looser than that of the hardened cement paste itself. In addition, the

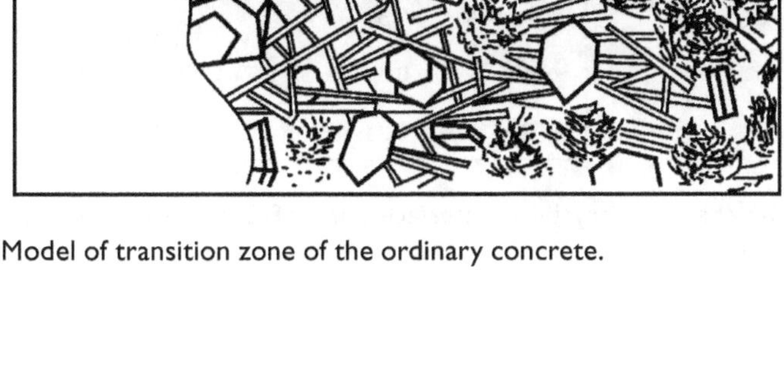

Figure 7.9 Model of transition zone of the ordinary concrete.

strength of the $Ca(OH)_2$ crystal is low, so the interface in ordinary concrete is a weak link and damage is often initiated from the interface. In the case of corrosion, failure starts from the interface.

In high strength high performance concrete, lowering the water/cement ratio along with addition of a superfine reactive mineral admixture leads to increased cohesion of the cement paste, the adsorbed water film on the aggregate becomes thinner, and the space for free crystalline growth diminishes. Besides, the reactive mineral admixture reacts with $Ca(OH)_2$ so the amount of CSH gel increases and the mass percentage of $Ca(OH)_2$ decreases. The crystallization of the hydration product is disturbed, therefore the particle size is decreased, and the degree of $Ca(OH)_2$ accumulation and orientation decreased. In addition, due to the filling effect of particles less than 1 μm, the transition zone becomes more compact making the interface dense and strong.

Zhongwei and Huizhen[10] studied the interfacial transition zone between the hardened cement paste and aggregate in mixes with water/cement ratios of 0.4 to 0.23 by X-ray diffraction. Figure 7.10 illustrates the variation in dimensions of $Ca(OH)_2$ crystals in the interfacial zone. Figure 7.11 illustrates the variations in dimensions of ettringite (Aft) crystals in the interfacial zone. Figure 7.12 shows the change in orientation of the $Ca(OH)_2$ crystal.

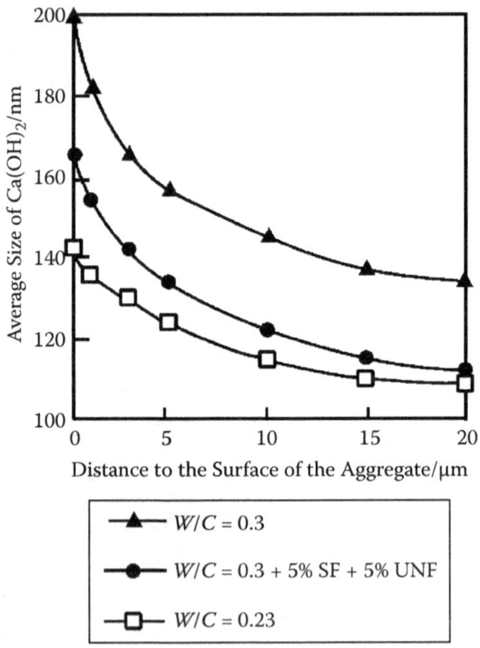

Figure 7.10 Dimension of crystal in interface zone of $Ca(OH)_2$ at low water/cement ratio.

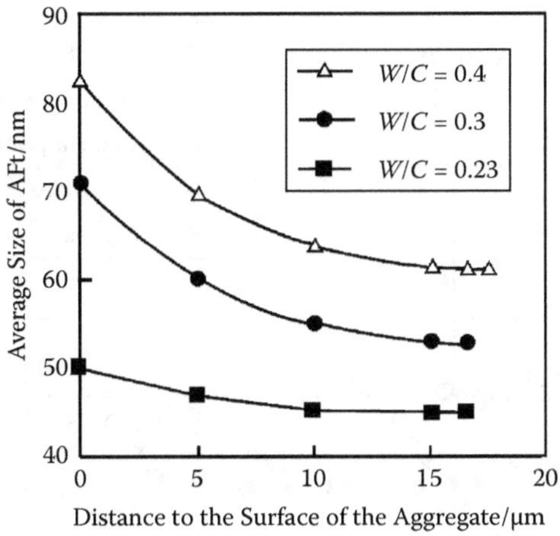

Figure 7.11 Variation of AFt crystal dimension at low water/cement ratio.

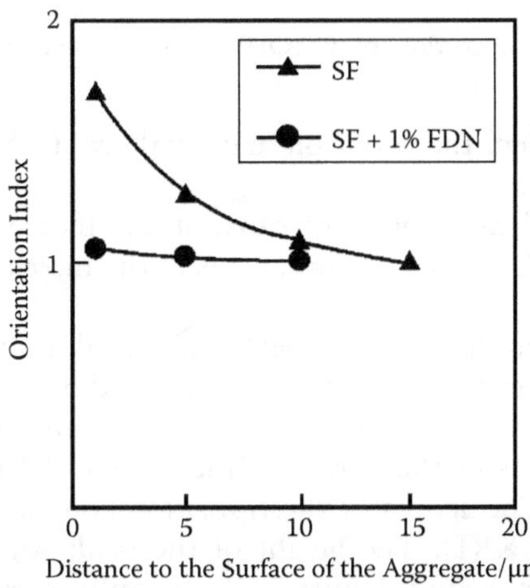

Figure 7.12 Influence of admixture on the orientation of $Ca(OH)_2$ in the interface zone of concrete.

It can be seen from Figures 7.10 and 7.11:

1. The closer the crystal grain lies to the interface, the larger the crystal dimension.
2. The dimensions of $Ca(OH)_2$ crystal grains are much larger than those of AFt crystals.
3. With a decrease in the water/cement ratio, dimensions of both $Ca(OH)_2$ and AFt crystals were decreased. The lower the water/cement ratio,

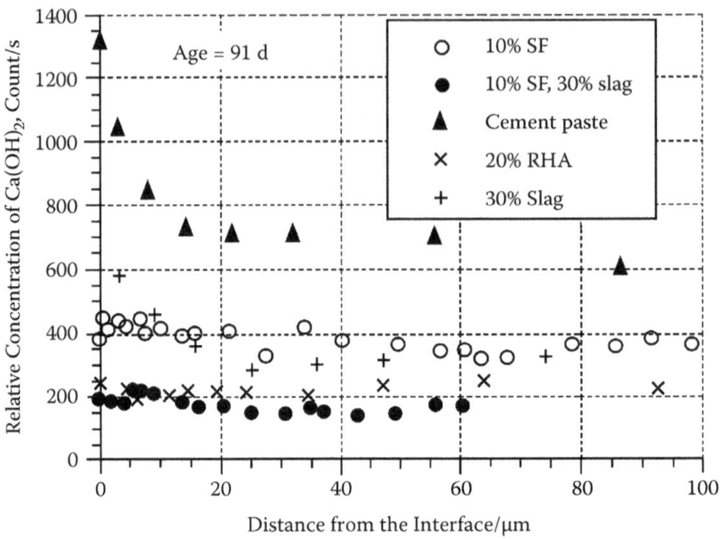

Figure 7.13 Relative concentration of $Ca(OH)_2$ in the interface zone (water/binder ratio 0.35).

the less dependent the variation in crystal grain size is on its distance from the interface.

4. Addition of silica fume may cause a significant reduction in the dimensions of $Ca(OH)_2$ crystal grains in the interfacial zone.

It can be seen from Figure 7.12 that with the addition of a superplasticizer, the degree of orientation of $Ca(OH)_2$ at 28 d diminished to zero.

Nilsen[60] has studied the influence of mineral admixture on the transition zone. The heights of the X-ray diffraction (XRD) peak at $2\theta = 34.1$ in concrete at different depths of the transition zone have been measured and compared with XRD. The height of the peak was expressed in the numbers counted per second (Figure 7.13). It can be seen from Figure 7.13 that at the age of 91 d, in the transition zone of concrete with an active mineral admixture, the relative concentration of $Ca(OH)_2$ is low, especially for cement paste containing 10% silica fume and 30% slag, where the concentration of $Ca(OH)_2$ is the least. For the neat cement paste without an additive, the concentration of $Ca(OH)_2$ was much higher than in the cement paste with an additive, and the closer to the interface, the higher the concentration. This proves that mineral admixture addition results in reduction of $Ca(OH)_2$ concentration.

The variations in microhardness across the interfacial zone of concrete determined by the microhardness tester may illustrate the properties of the interface zone. Figure 7.14 shows the characteristics of microhardness

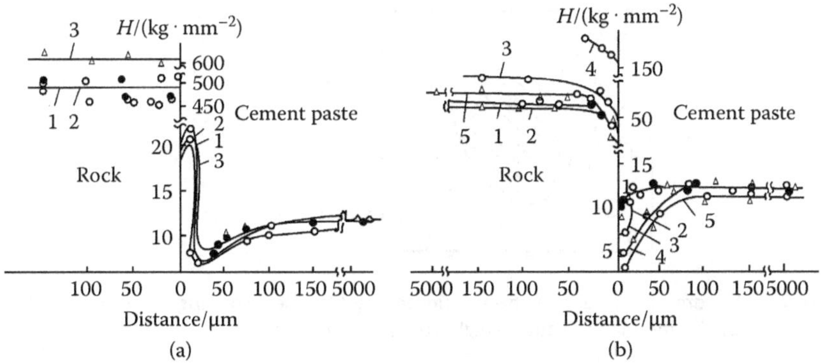

Figure 7.14 Distribution of microhardness of the interface transition zone of ordinary concrete. (a) Acidic aggregate: 1 = Plagioclase, 2 = Basalt, 3 = Quartzite; (b) Carbonate aggregate: 1 = Limestone, 2 = Calcite (parallel to cleavage), 3 = Marble, 4 = Dolomite, 5 = Calcite (perpendicular to cleavage). (From Chongqing Institute of Architecture and Civil Engineering and Nanjing Institute of Technology, *Science of Concretes*, Beijing: China Building Industry Press, 1983 [in Chinese].)

variation in the interface zone of ordinary concrete. It can be seen that whichever aggregate is used, there is a valley of microhardness in the interface transition zone. Obviously, this is the weak link in the interface zone of ordinary concrete. Chongqing Institute of Architecture and Civil Engineering and Nanjing Institute of Technology[19] showed that in ordinary heterogeneous concrete, the bond in the interface between the coarse aggregate and the mortar was the weak link, with the strength of the bond between the coarse aggregate and the mortar being 35% to 65% of that found within the mortar.

Wang Yongwei[57] studied the microhardness of the interfacial zone of the cement paste with the aggregate in super-high-strength high performance concrete. The cement paste given in Table 7.3 was poured into a 2 cm × 2 cm × 2 cm specimen mold and a piece of limestone grain with a particle size around 5 mm was inserted in the center, then compacted by vibration. Special attention was paid to remove air bubbles around the aggregate grains and in the cement paste itself. The mold was removed after 1 d and the specimen was cured in the standard curing room for 56 d before being taken out and immersed in anhydrous alcohol for 1 d to stop the hydration, dried at 60°C to constant weight, and then ground to expose the aggregate, and the surface was polished for testing (Figure 7.15). Testing was carried out using an automatic microhardness tester made in Germany. Test results are shown in Figure 7.16 from where it can be seen that

1. In comparison with the microhardness of the interfacial transition zone of ordinary concrete, there is no valley in the curve of microhardness distribution across the interface in super-high-strength high

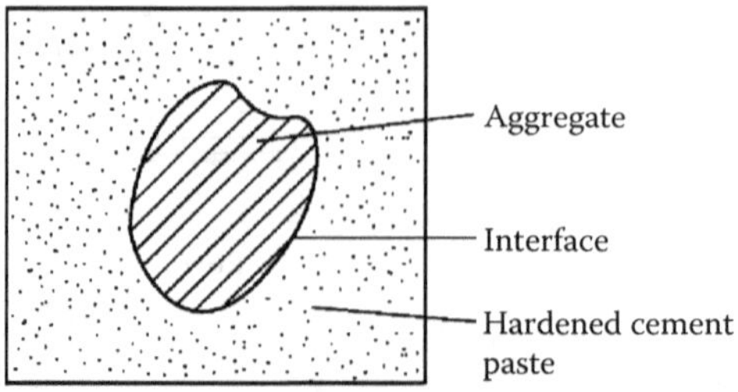

Figure 7.15 Schematic picture of the specimen used for determining the microhardness of the interface in super-high-strength concrete.

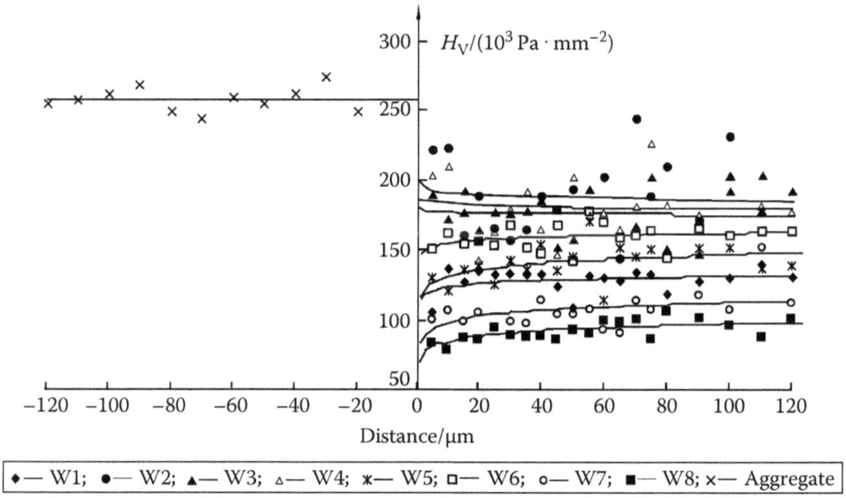

Figure 7.16 Test results of microhardness of interface in super-high-strength high performance concrete.

performance concrete. The structure of the interface was improved greatly with the microhardness of the interface zone slightly lower than that of cement paste itself. For example, the samples without mineral admixture (W1) and with a water/binder ratio of 0.25 (W7), with a water/binder ratio of 0.30 (W8), W6 with single addition of silica fume, and W5 with expanding admixture show little difference between them. But for the mixes with a water/binder ratio of 0.20 and addition of silica fume plus slag (W2), addition of silica fume and fly ash (W3), and that with silica fume and X powder (W4), the microhardness of the interfacial zone was higher than that of the cement paste itself showing that the interface was not weakened but strengthened, demonstrating that improvement of the structure and property of the interfacial zone is possible. It can be seen from Table 7.4 that the 56 d compressive strengths of concretes from these mixes were also the highest and reached 146.0, 134.7, and 136.1 MPa, respectively.

2. When the microhardness of limestone aggregate was compared with that of the cement paste, the difference was smaller than that in ordinary concrete, illustrating that the inner structure of super-high-strength high performance concrete tends to be homogeneous.
3. At the same water/binder ratio of 0.2, the microhardness of a cement paste containing an active mineral admixture is higher than that of neat cement paste consistent with the increase in compressive strength.
4. Comparing the single addition of 10% silica fume (W6) with mixes containing combined addition, such as 10% silica fume plus 20% slag (W2), 10% silica fume plus 20% fly ash (W3), and 10% silica fume plus 20% X powder (W4), the microhardness of the interface was higher, and higher than that of the cement paste itself. Such a strengthening effect is the result of the combined influence of filling and compacting.
5. Comparing mixes W6, W7, and W8, and the increase in water/binder ratio, the microhardness of both the interface and cement paste was lowered and all were slightly lower than that of the cement paste itself. Similarly, the compressive strength at 56 d in Table 7.4 also showed that with an increase of water/binder ratio, the strength decreased.
6. The microhardness of a cement paste with a combined addition of 10% silica fume and 12% UEA expansion admixture (W5) was lower than that of cement paste with a single addition of 10% silica fume, obviously related to the effect of the expansive admixture.

Except for the study of the interface zone of super-high-strength high performance concrete using microhardness, the structure of the interface in super-high-strength high performance concrete has been observed by scanning electron microscope (SEM). In that study, 100 mm × 100 mm × 100 mm concrete cubes cured for 56 days under standard conditions were examined after the compressive strengths were determined. Broken pieces from the center of the specimen were immersed in anhydrous alcohol, to stop hydration, dried to constant weight at 60°C, then a piece with an interface of cement paste with aggregate was selected, broken into a 1-cm^3 piece and surfaced with gold, and examined by SEM.

Figure 7.17(a) is a SEM micrograph of the interfacial zone of super-high-strength concrete W11 without an active mineral admixture at water/binder ratio 0.2 magnified 2000 times. It can be seen that the interfacial zone is dense without an apparent transition zone although there are large amounts of needle-like crystals near the interface. Figure 7.17(b) is a SEM photo of the interfacial zone of super-high-strength concrete W12 containing 10% silica fume and 20% slag at a water/binder ratio of 0.2 magnified

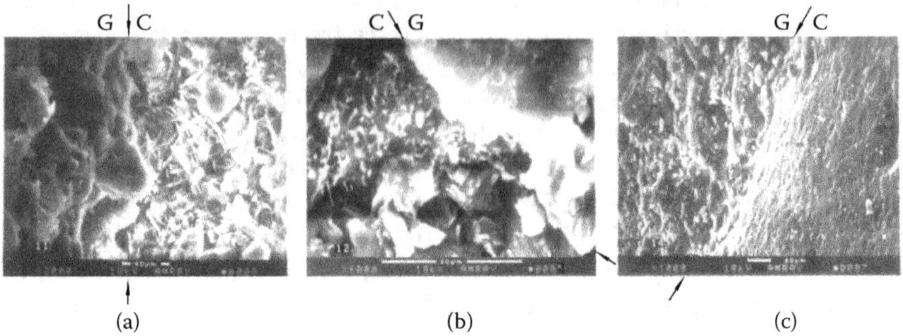

Figure 7.17 SEM photo of interfacial zone of super-high-strength high performance concrete. (G = Aggregate, C = Hardened cement paste.)

by 6000 times, where the cement paste in the interface zone is even denser. Figure 7.17(c) is a SEM photo of the interfacial zone of super-high-strength concrete W14 with 10% silica fume and 20% X powder, magnified 1000 times. Like the W12 sample, there is no obvious pore space nor or there micro-openings and the interfacial zone is dense. All of the SEM photos of the super-high-strength concrete mixes given in Table 7.4 showed excellent compactness of the transition zone (see Wang Yongwei[54]).

It can be seen from the discussion above that the orientation index of $Ca(OH)_2$ is reduced within the interface transition zone of concrete when superplasticizer is used at a low water/binder ratio. The concentration of $Ca(OH)_2$ was reduced, with low porosity and high density. When the water/binder ratio was reduced to 0.2, the microhardness of the interfacial transition zone of concrete with an active mineral admixture exceeded that of the cement paste itself. In that case, the interfacial zone was no longer the weak link in the concrete. This lays the foundation for the excellent physico-mechanical properties of super-high-strength high performance concrete.

7.4 MACROSTRUCTURE OF SUPER-HIGH-STRENGTH HIGH PERFORMANCE CONCRETE

Concrete is an artificial stone made from binding coarse and fine aggregates into a monolith with a hydrated cement paste; its macrostructure is an agglomerate. In ordinary concrete, due to both the low strength and modulus of elasticity of the cement paste, the bond of the interface is weak and the aggregate is a dense and strong grain, so the heterogeneity of the concrete is quite apparent.

In super-high-strength high performance concrete, the situation changes considerably. The strength of the cement paste is enhanced greatly, to almost the strength of the aggregate; the compactness and the strength of the interfacial zone between cement and aggregate are enhanced and improved greatly and even exceeding those of the cement paste itself. In such a case, the homogeneity of super-high-strength high performance concrete is enhanced and close to homogeneous material.

The author has carried out comparison measurements on neat paste, mortar, and concrete. The cement used was Tenghui 525 OPC, silica fume from Qinzhen, medium sand from Jianyang (fineness modulus 2.4), limestone from Xiaoquan, Chongqing City (maximum size 20 mm), and melamine superplasticizer. The basic parameters of the formulations for the three different types of specimens were about the same and the dimensions were 100 mm × 100 mm × 100 mm with test results given in Table 7.6. It can be seen from this table that the quality of the hardened paste in the mortar is the same as that in the neat paste, and the quality of the mortar in the concrete is the same as that of the mortar specimen, while the quality of the hardened paste in the concrete is the same as that in both mortar and neat paste. With the introduction of fine and coarse aggregates, the fluidity of the mortar and the concrete mix is decreased. The 28 d compressive strength reaches around 140 MPa (the small difference is due to experimental error). From the viewpoint of the strength of material, after introduction of fine and coarse aggregates, the strength of material remains unchanged. Because the density of sand and limestone is higher than that of hardened cement paste, introducing fine and coarse aggregates increases the apparent density of the material.

Due to the very high strength of the cement paste in super-high-strength high performance concrete and the strong interfacial bond, during failure under loading, the rupture path passes through aggregate and cement paste, and not along the interface. The fracture path is straight, not curved, and no loose aggregate particles lie on the broken surface. If a specimen of super-high-strength concrete is split into two halves (Figure 7.18), the particles of aggregate are broken into two symmetrical parts.

According to rupture theory, when a specimen is broken under loading, the crack is initiated from the weakest point of the material and spreads along the path with the minimum energy consumption. When the crack encounters an aggregate particle, the issue is whether the crack will propagate along the interface between cement paste and aggregate or pass directly through the aggregate.

This depends on the energy consumed along the path. As shown in Figure 7.19, assuming that path 1 passes directly through the aggregate while path 2 is along the interface, S_1 and S_2 are the lengths of the path 1 and path 2, respectively, γ_1 and γ_2 are the energies of rupture per unit length of aggregate and interface; then if $S_1\gamma_1 > S_2\gamma_2$, the path of rupture is along the interface, while when $S_1\gamma_1 < S_2\gamma_2$, the path of rupture passes through the aggregate. In general, aggregate strength is high, so γ_1 is large, whereas in

Table 7.6 Comparative test of super-high-strength paste, mortar, and concrete

Kind of specimen	Parameters of mix					Fluidity (mm)		28 d ρ_a (kg/m^3)	Stardand curing, 28 d R_{com} (MPa)
	W/B	Silica fume/%	SuperPlasticizer (%)	Paste/sand ratio	Mortar/coarse aggregate	Slump	Spread		
Neat paste	0.16	10	2	—	—	284	820	2332	141.4
Mortar	0.16	10	2	1.144	—	255	563	2445	138.6
Concrete	0.16	10	2	1.144	1.430	141	—	2554	140.3

Note: W/B = water/binder ratio; ρ_a = apparent density; R_{com} = compressive strength.

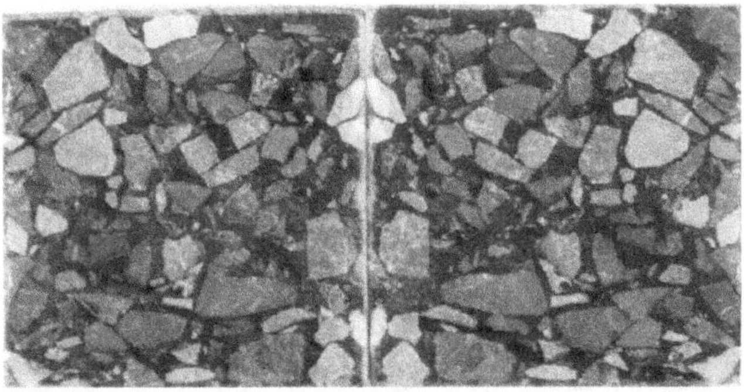

Figure 7.18 Macrostructure of super-high-strength high performance concrete. (Dimension of the specimen: 100 mm ×100 mm ×100 mm.)

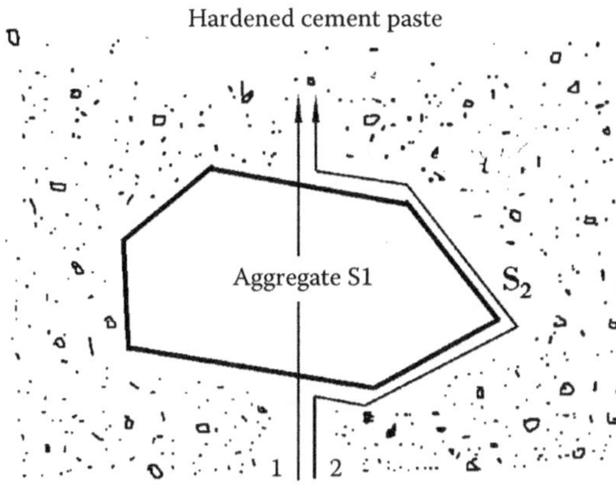

Figure 7.19 Schematic rupture path of the concrete.

ordinary concrete, and in particular gravel concrete, due to the weak bond of the interface, γ_2 is low, so although $S_2 > S_1$, $S_2\gamma_2$ is always less than $S_1\gamma_1$. Therefore, the path of rupture goes along the interface.

In super-high-strength concrete, the strength of the interfacial transition zone (ITZ) has greatly increased. As shown in Figure 7.16, the microhardness of the ITZ is close to that of the aggregate and γ_2 is significantly improved, and S_2 is more than S_1. As a result, $S_2\gamma_2$ is always more than $S_1\gamma_1$, so therefore the path of rupture goes through the aggregate demonstrating the superiority and hardness of super-high-strength high performance concrete.

Chapter 8

Strength and deformation properties

At present, both in China and abroad, there has been little study on the strength and deformation properties of super-high-strength high performance concrete. What has been conducted has investigated the strength or single deformation indices of the lower class of super-high-strength high performance concrete. Consequently, the data are limited and scattered. Today, only in Norwegian and German standards are there specifications for the lower class of concrete in the super-high-strength concrete category. In the Norwegian standard, NS 3473E, the highest strength class is C105, while in the *Guide to High Strength Concrete* issued in 1995 by the German Concrete Association, the highest strength grade is C115.

To promote the application of super-high-strength high performance concrete in future engineering, the author and his colleagues have systematically studied its strength properties, deformation under short period loading, and shrinkage.

8.1 THE STRENGTH PROPERTIES OF SUPER-HIGH-STRENGTH HIGH PERFORMANCE CONCRETE

The axial compressive strength, splitting strength, flexural strength, and bond strength of concrete with steel reinforcement were studied and the correlation of these with compressive strength has been determined.

8.1.1 Axial compressive strength

The axial compressive strength of super-high-strength high performance concrete was obtained on 100 mm × 100 mm × 300 mm prism specimens. Twenty-four groups (three specimens for each group) were tested, with the results given in Table 8.1 and Figure 8.1. It can be seen from the data in Table 8.1 and from Figure 8.1 that as the compressive strength of concrete (100 mm × 100 mm × 100 mm specimen) increased, the axial compressive strength increased. Though the specimen preparation and the tests were

Table 8.1 Axial compressive strength of super-high-strength concrete

No.	Compressive strength $f_{cu,10}$ (MPa)	Axial strength $f_{c,10}$ (MPa)	$f_{c,10}/f_{cu,10}$	No.	Compressive strength $f_{cu,10}$ (MPa)	Axial strength $f_{c,10}$ (MPa)	$f_{c,10}/f_{cu,10}$
G1	98.3	90.913	0.93	G13	118.1	98.185	0.83
G2	100.0	95.786	0.96	G14	120.1	104.840	0.87
G3	100.5	93.148	0.93	G15	121.0	109.095	0.90
G4	102.9	81.531	0.79	G16	121.3	107.185	0.88
G5	102.9	80.268	0.78	G17	122.1	107.432	0.88
G6	108.0	96.129	0.89	G18	125.3	90.963	0.73
G7	110.0	101.134	0.92	G19	131.6	101.803	0.77
G8	114.0	102.741	0.90	G20	133.4	98.222	0.74
G9	114.8	92.765	0.81	G21	134.5	108.327	0.81
G10	115.0	90.050	0.78	G22	139.7	111.012	0.79
G11	116.0	101.333	0.87	G23	140.1	110.333	0.79
G12	117.6	107.358	0.91	G24	148.6	123.152	0.84

Note: For samples G2, G6, G10, G12, and G14, the coarse aggregate is basalt, while for the other groups it is limestone.

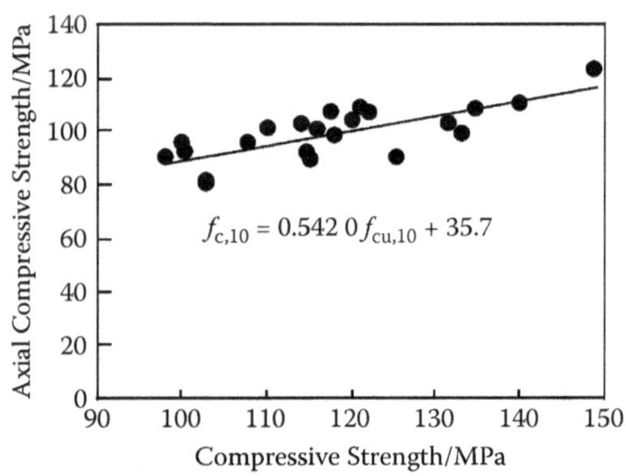

Figure 8.1 Relationship between axial compressive strength and compressive strength of super-high-strength concrete.

carried out carefully, the ratios of axial compressive to cubic compressive strength results are scattered, with a maximum of 0.96 and a minimum of 0.73 with an average of 0.85. The reason is that the strength of the concrete is high, so ensuring the two side surfaces are parallel, together with the uneven distribution of initial defects within the specimen and the difficulty in ensuring geometrical and physical centering during the test, all have an influence on the results.

According to the data of the 19 groups of super-high-strength concretes made with limestone coarse aggregate shown in Table 8.1, the equation of the relationship between axial compressive strength and compressive strength of super-high-strength concretes with limestone coarse aggregate by regression is

$$f_{c,10} = 0.542\,0\, f_{cu,10} + 35.7 \tag{8.1}$$

$$100 \text{ MPa} \le f_{cu,10} \le 150\text{MPa}$$

8.1.2 Splitting strength

Considering the difficulty of conducting the axial tensile test of concrete and the scattered results obtained, we used only the splitting test on specimens with dimensions 100 mm × 100 mm × 100 mm, which is the same as for the compressive strength, and followed the Chinese Standard, GBJ 81-85. Seventeen groups of specimens were tested (three specimens for each group) and test results are given in Table 8.2 and in Figure 8.2. It can be seen from Figure 8.2 that with an increase of compressive strength of super-high-strength concrete, the splitting strength was also increased but at a lower rate. The relationship between splitting strength ($f_{pl,10}$) and cubic compressive strength ($f_{cu,10}$) of super-high-strength concrete is as follows:

$$f_{pl,10} = 0.040\,3\, f_{cu,10} + 2.26 \tag{8.2}$$

$$90 \text{ MPa} \le f_{cu,10} \le 150 \text{ MPa}$$

The ratio of splitting strength to cubic compressive strength calculated from the data given in Table 8.2 varied between 1/14.6 and 1/19.6, which

Table 8.2 Test results of splitting compressive strength of super-high-strength high performance concrete

No.	*Compressive strength (MPa)*	*Splitting strength (MPa)*	*Tension/ compression ratio*	No.	*Compressive strength (MPa)*	*Splitting strength (MPa)*	*Tension/ compression ratio*
1	92.1	5.90	1/15.6	10	121.3	6.84	1/17.7
2	96.7	5.98	1/16.8	11	122.1	6.18	1/19.6
3	100.5	6.54	1/15.7	12	125.3	7.32	1/17.1
4	102.9	7.41	1/14.6	13	132.4	7.13	1/18.6
5	108.0	6.14	1/17.9	14	134.5	7.18	1/18.7
6	109.8	6.71	1/17.0	15	139.7	8.13	1/17.2
7	113.8	7.77	1/14.9	16	140.1	7.62	1/18.4
8	116.1	7.19	1/16.4	17	148.6	8.82	1/16.8
9	117.6	7.96	1/15.2				

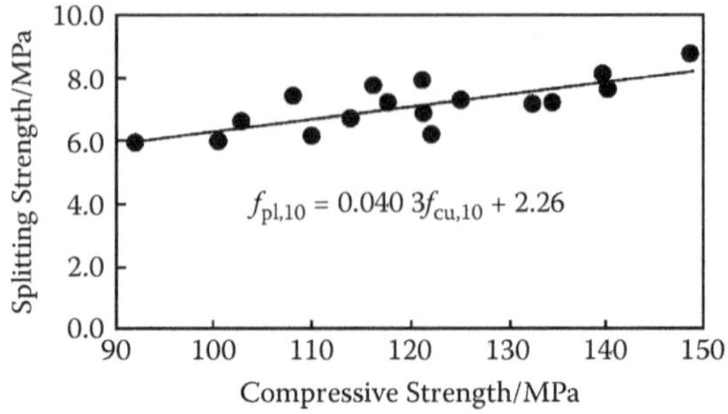

Figure 8.2 Relationship between splitting strength and compressive strength of super-high-strength concrete.

is much lower than the tension/compression ratio (usually about 1/10) of ordinary concrete, illustrating the much higher brittleness of super-high-strength concrete compared to ordinary concrete. In fact, failure of super-high-strength concrete under compression is like a blast in the way the sample shatters.

8.1.3 Flexural strength

The specimen for the flexural test was a concrete beam of dimensions 100 mm × 100 mm × 400 mm, loaded at third points by two equal concentrated loads; the span between the supports was 300 mm (based on China Standard GBJ 81-85). Eighteen groups of specimens were tested (three specimens for each group), with test results given in Table 8.3 and in Figure 8.3. It can be seen from Figure 8.3 that with the increase

Table 8.3 Test results of flexural strength of super-high-strength high performance concrete

No.	Compressive strength (MPa)	Flexural strength (MPa)	Flexural/ compression ratio	No.	Compressive strength (MPa)	Flexural strength (MPa)	Flexural/ compression ratio
1	92.1	10.41	1/8.8	10	121.1	13.62	1/8.9
2	96.7	10.13	1/9.5	11	121.3	14.07	1/8.6
3	100.5	13.14	1/7.6	12	125.3	12.72	1/9.9
4	102.9	11.31	1/9.1	13	131.6	15.78	1/8.3
5	108.0	10.18	1/10.6	14	132.4	14.85	1/8.9
6	109.8	10.64	1/10.3	15	134.5	13.79	1/9.8
7	113.8	12.55	1/9.1	16	139.7	13.77	1/10.1
8	116.1	14.86	1/7.8	17	140.1	15.72	1/8.9
9	117.6	11.61	1/10.1	18	148.6	15.67	1/9.5

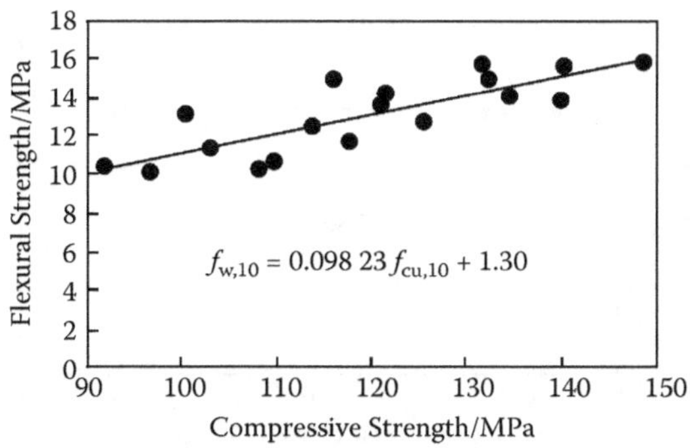

Figure 8.3 Relationship between flexural strength and compressive strength of super-high-strength concrete.

of compressive strength of super-high-strength concrete, the flexural strength is increased as well but at a lower rate. Based on the test data in Table 8.3, the flexural/compression ratio is 1/7.6 to 1/10.6. The relationship between flexural strength ($f_{w,10}$) and compressive strength is as follows:

$$f_{w,10} = 0.098\ 23\ f_{cu,10} + 1.30 \tag{8.3}$$

$$90\ \text{MPa} \le f_{cu,10} \le 150\ \text{MPa}$$

8.1.4 Bond strength of the steel reinforcement with super-high-strength concrete

In the study, the bond strength to a 16-mm-diameter round plain carbon steel bar No. 3 was used. Specimens were made up to dimensions of 100 mm × 100 mm × 200 mm with the steel bar embedded in the center of the specimen to a depth of 200 mm. The bond strength of each mix was the mean value from the results of six specimens, and is given in Table 8.4 and Figure 8.4.

Table 8.4 Test results of bond strength of steel bar with super-high-strength concrete

No.	*Compressive strength (MPa)*	*Bond strength (MPa)*	No.	*Compressive strength (MPa)*	*Bond strength (MPa)*	No.	*Compressive strength (MPa)*	*Bond strength (MPa)*
1	92.1	5.83	4	110.1	5.70	7	139.7	5.98
2	96.7	6.00	5	121.3	6.50	8	140.1	5.93
3	102.9	5.82	6	134.5	6.00	9	148.6	6.39

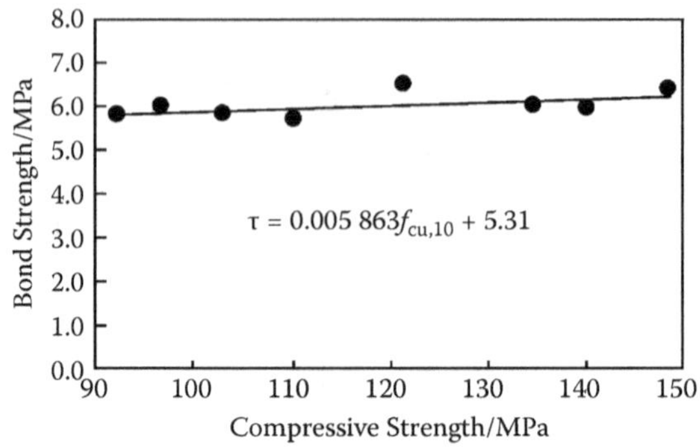

Figure 8.4 Relationship between bond strength of round plain steel bar with super-high-strength concrete and compressive strength of concrete.

It can be seen from Figure 8.4 that with an increase in the compressive strength of super-high-strength concrete, the bond strength of concrete to a round plain steel bar (τ) was increased only slightly. Consequently, it can be concluded that the influence of compressive strength on the bond strength of super-high-strength concrete with round plain steel bar is not significant and based on the data given in Table 8.4, the following equation can be developed:

$$\tau = 0.005\ 863\ f_{cu,10} + 5.31 \tag{8.4}$$

$$90\text{ MPa} \le f_{cu,10} \le 150\text{ MPa}$$

8.2 DEFORMATION OF SUPER-HIGH-STRENGTH HIGH PERFORMANCE CONCRETE UNDER SHORT-TERM LOADING

8.2.1 Stress–strain curve

To determine the stress–strain curve precisely along with other key physico-mechanical parameters, two methods, strain gauge and concrete strain foil, were applied to measure longitudinal and lateral strain of the specimen. The longitudinal and lateral gauge lengths were 200 mm and 90 mm, respectively, while those for the strain foil were 150 mm and 80 mm, respectively, in the arrangement shown in Figure 8.5. The concrete strain foil was observed during the test and was not damaged until the failure of the concrete so it is an ideal tool for the test.

Specimen dimensions for determining stress–strain curve were 100 mm × 100 mm × 300 mm. Testing was carried out on a 200-tonne INSTRON electro-servo pressure machine in the geotechnical laboratory of Chongqing

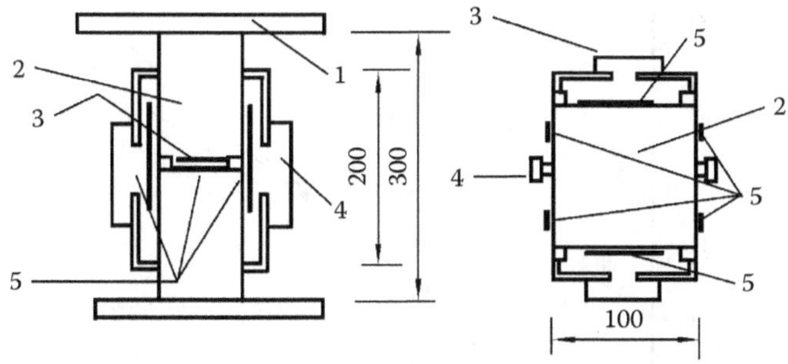

Figure 8.5 Arrangement of strain measuring instruments on super-high-strength concrete: 1 = plate; 2 = specimen; 3 = lateral strain gauge; 4 = longitudinal strain gauge; 5 = strain foil.

University. The loading was controlled by displacement with the axial pressure controlled automatically at a loading rate of 5×10^{-6} kN/s. In order to record the axial pressure simultaneously, a 2000-kN pressure sensor was added under the specimen, and during the test the axial pressure and the longitudinal and lateral strains were recorded throughout the whole process. The data from the strain gauge were plotted using the X-Y function meter of the INSTRON machine and the other data were collected automatically by the computer.

Due to the super high strength of the specimen, all specimens broke suddenly and shattered. Before failure, there was no visible crack, although for some specimens there was a cracking sound inside the specimen just before the moment of failure. At this point, the stress–strain curve began to level out and suddenly fell off after the specimen had split and burst. Only the upper and lower cones were left, with the ruptured surface plain and smooth, passing through the aggregate. (A container was placed outside the specimen to avoid harm from the fragments.) Test results showed that the descending part of the stress–strain curve could not be obtained even if the closed loop of the electric servo controlled by displacement was applied. The stress–strain curve is shown in Figure 8.6.

It can be seen from Figure 8.6 that the stress–strain curve of super-high-strength concrete is characterized by the following features:

1. The rising part of the stress–strain curve is close to a straight line, even in the period close to damage, so the longitudinal stiffness goes through no obvious change.
2. With increasing strength of the concrete, the peak strain of the concrete increased significantly.
3. The ultimate stress and peak stress coincide, as well as do the ultimate strain and peak strain.

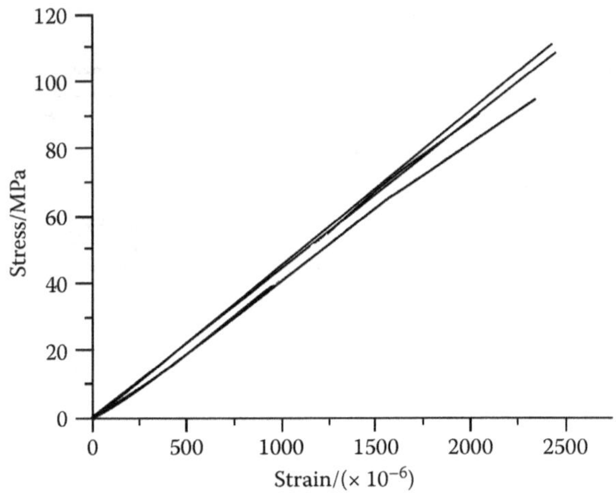

Figure 8.6 Stress–strain relationship of super-high-strength concrete.

8.2.2 Peak strain

Using the data in Table 8.5, an equation of the relationship between peak strain and compressive strength of limestone super-high-strength concrete can be obtained as follows:

$$\varepsilon_0 = (3.701\ 2\ f_{cu,10} + 1\ 913) \times 10^{-6} \quad (8.5)$$

$$90\ \text{MPa} \leq f_{cu,10} \leq 150\ \text{MPa}$$

The equation for the relationship between the peak strain and prism strength is as follows:

$$\varepsilon_0 = (8.091\ 9\ f_{c,10} + 1\ 553) \times 10^{-6} \quad (8.6)$$

It can be seen from Figure 8.7 that with an increase in the compressive strength of concrete, the peak strain value of super-high-strength concrete is raised only slightly.

8.2.3 Lateral deformation and Poisson's ratio

In Figure 8.8, the lateral and longitudinal strain curves for several specimens over the whole test are given. It can be seen that these curves are bent slightly in the direction of the ordinate, which illustrates that Poisson's ratio for super-high-strength concrete does increase during the test but only to a limited degree. Poisson's ratio values in Table 8.5 are

Table 8.5 Peak strain, Poisson ratio, and deformation modulus of super-high-strength high performance concrete

No.	*Compressive strength (MPa)*	*Peak strain* $\times 10^{-6}$	*Poisson's ratio*	*Deformation modulus (MPa)*	*No.*	*Compressive strength (MPa)*	*Peak strain* $\times 10^{-6}$	*Poisson's ratio*	*Deformation modulus (MPa)*
G1	98.3	2416	0.228	41,813	G13	118.1	2510	0.244	41,838
G2	100.0	2868	0.211	38,890	G14	120.1	2747	0.251	41,402
G3	100.5	2549	0.242	40,460	G15	121.0	2617	0.291	48,872
G4	102.9	2115	0.225	41,689	G16	121.3	2314	0.221	46,675
G5	102.9	2160	0.220	43,203	G17	122.1	2304	0.184	48,017
G6	108.0	2690	0.249	43,841	G18	125.3	2440	0.241	39,672
G7	110.0	2415	0.275	48,952	G19	131.6	2637	0.264	42,742
G8	114.3	2245	0.187	46,712	G20	133.4	2278	0.290	45,741
G9	114.8	1934	0.232	46,344	G21	134.5	2366	0.235	48,725
G10	115.0	2645	0.218	40,765	G22	139.7	2305	0.242	49,650
G11	116.0	2265	0.275	50,399	G23	140.1	2378	0.199	49,497
G12	117.6	2781	0.232	43,358	G24	148.6	2597	0.247	50,291

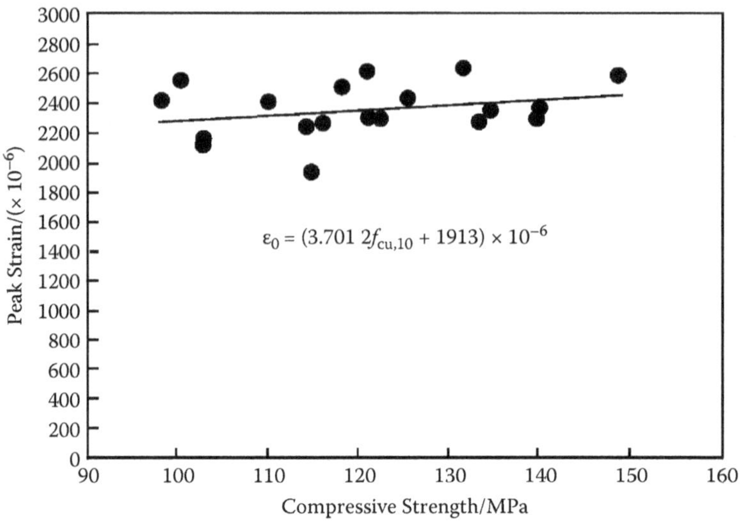

Figure 8.7 Relationship between peak strain and compressive strength.

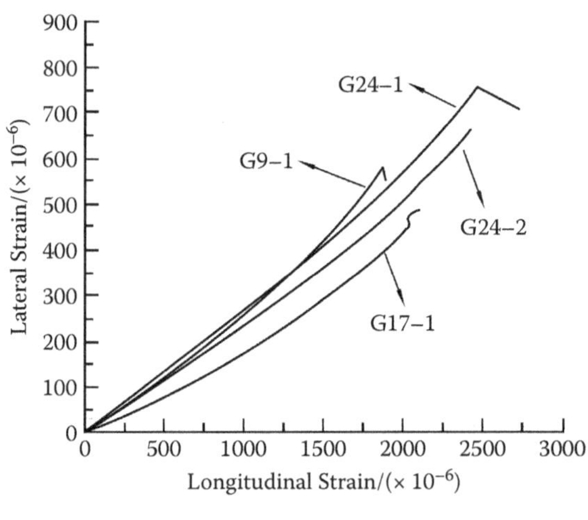

Figure 8.8 Relationship between lateral and longitudinal strains.

the mean values calculated from the results based on the data of lateral and longitudinal deformations with the stress of 0.2 $f_{c,10}$ to 0.4 $f_{c,10}$. Figure 8.9 shows the relationship between Poisson's ratio and compressive strength of super-high-strength concrete where it can be seen that Poisson's ratio values are scattered with an average of 0.238. The ACI

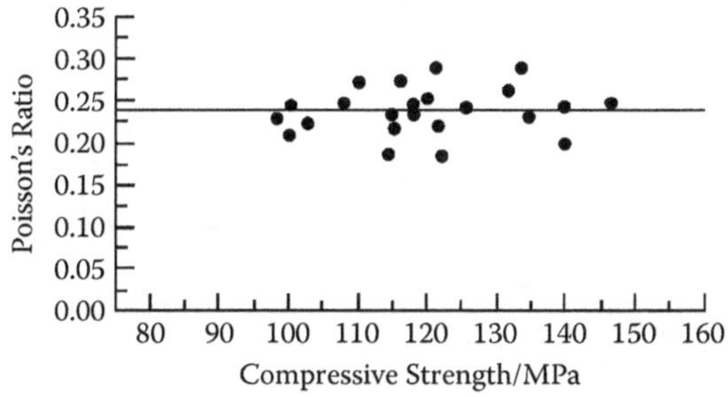

Figure 8.9 Relationship between Poisson's ratio and compressive strength of super-high-strength concrete.

high strength concrete committee reported that the test results for the ratio were 0.20 to 0.28 for 50- to 80-MPa high strength concretes, while Poisson's ratio for concrete with compressive strengths of 63.9 MPa and 102.0 MPa were determined by the China Railway Academy as 0.22 and 0.23, respectively.[61] In general, Poisson's ratio of super-high-strength concrete is slightly higher than that of ordinary concrete and close to that of high strength concrete.

8.2.4 Deformation modulus

The stress–strain curve of super-high-strength concrete is close to a straight line. In this study, the deformation modulus was the mean calculated value of the results based on the stress in the range of 0.2 $f_{c,10}$ to 0.4 $f_{c,10}$ and related strain. The results are given in Table 8.5 and in Figure 8.10. It can be seen from the figure that with an increase in the compressive strength of concrete, the modulus of deformation increased. From non-linear fitting of the data from our study, the equation for the relationship between deformation modulus (E_0) and compressive strength ($f_{cu,10}$) of limestone super-high-strength concrete can be derived:

$$E_0 = (0.287\sqrt{f_{cu,10}} + 1.438) \times 10^4 \tag{8.7}$$

Curve 1 in Figure 8.10 is based on the equation for ordinary concrete given in China Specification GBJ 10-89,

$$E_0 = \frac{10^5}{2.2 + \dfrac{34.5}{f_{cu,10}}} \tag{8.8}$$

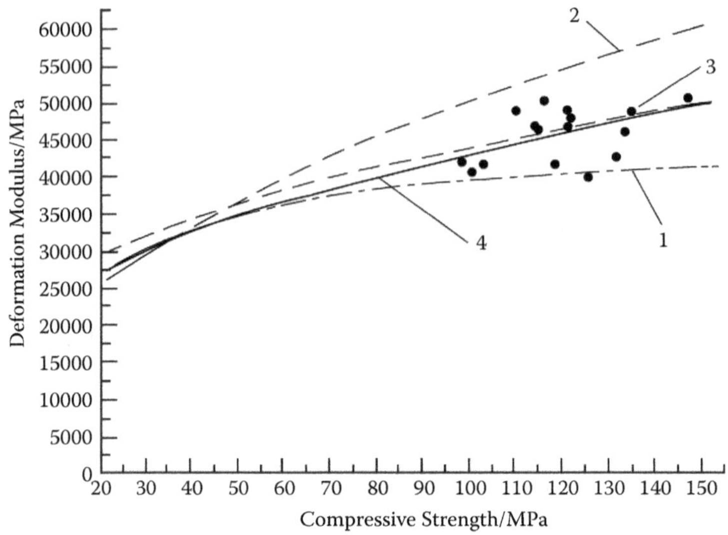

Figure 8.10 Relationship between deformation modulus and compressive strength of super-high-strength concrete.

Curve 2 is calculated using the equation for the deformation modulus of high strength concrete presented by Chen Zaoyuan, Zhu Jinquan, and Wu Peigang[61] where

$$E_0 = (0.45\sqrt{f_{cu,10}} + 0.5) \times 10^4 \tag{8.9}$$

Curve 3 is calculated using the empirical equation for the deformation modulus of high strength concrete presented by Xu Jin-feng,[62] where

$$E_0 = (0.261\sqrt{f_{cu,10}} + 1.787) \times 10^4 \tag{8.10}$$

It can be seen from Figure 8.10 that curves 1 and 2 define the upper and lower limits of the data from the study by the author and his colleagues. The fitted equation (curve 4) suggested by the author and his colleagues for high strength concretes is close to that of curve 3, while the calculated results for low strength concrete (<50 MPa) lie close to curve 1, meaning it can serve as the general equation for determining the deformation modulus of ordinary, high strength, and super-high-strength concretes.

8.3 MECHANICAL PROPERTIES INDICES OF SUPER-HIGH-STRENGTH HIGH PERFORMANCE CONCRETE

Based on the equations for the relationship of strength indices, peak strain, and deformation with the compressive strength of super-high-strength concrete obtained by the author and his colleagues, we can calculate the mechanical property indices for a series of super-high-strength concretes with different strengths, providing the basic data to understand and apply super-high-strength concrete.

From the results above (Table 8.6), the mechanical properties of super-high-strength concrete can be characterized as follows:

1. The splitting and flexural strengths of super-high-strength concrete increase with an increase of compressive strength but the rate of increase is lower than that of ordinary concrete, further illustrating the brittleness of super-high-strength concrete.
2. The rising part of the stress–strain curve for super-high-strength concrete is basically a straight line. The stress at the limit of proportionality is close to the peak stress with Poisson's ratio less than 0.5. There is no critical stress point at which cracking occurs, total failure being close to peak stress. The longitudinal deformation is basically the elastic compressive deformation with little or no plastic deformation. When the stress reaches its peak value, the strength of the cement paste is approximately equal to that of the coarse aggregate, which cannot prevent cracking. At failure, a large amount of the energy accumulated inside super-high-strength concrete is released rapidly and the sample shatters so the descending part of the stress–strain curve cannot be measured.
3. The peak strain of super-high-strength concrete increases with an increase in compressive strength and coincides with the ultimate strain. The curves of lateral and longitudinal deformation of super-high-strength concrete maintain their linear relationship at initial and medium stages with only a non-linear one in latter stages (after 0.6 $f_{cu,10}$) with little bending.

The equations derived for the relationship of strength and deformation properties with compressive strength and the mechanical property indices of super-high-strength concrete of different grades offer basic data for the application of super-high-strength high performance concrete.

Table 8.6 Mechanical property indices of super-high-strength concretes

		Grade of super-high-strength concrete					
Index	*Equation*	*C100*	*C110*	*C120*	*C130*	*C140*	*C150*
$f_{c,10}$·(MPa)	$f_{c,10} = 0.542\, f_{cu,10} + 35.7$	89.9	95.3	100.7	106.2	111.6	117.0
$f_{w,10}$·(MPa)	$f_{w,10} = 0.098\,23\, f_{cu,10} + 1.3$	11.12	12.11	13.09	14.07	15.05	16.03
$f_{pl,10}$·(MPa)	$f_{pl,10} = 0.040\,32\, f_{cu,10} + 2.26$	6.29	6.70	7.10	7.50	7.90	8.31
τ (MPa)	$\tau = 0.005\,863\, f_{cu,10} + 5.31$	5.90	5.95	6.01	6.07	6.13	6.19
ε_0	$\varepsilon_0 = (3.7012\, f_{cu,10} + 1913) \times 10^{-6}$	2283	2320	2357	2394	2431	2468
E_0·(MPa)	$E_0 = \left(0.287\sqrt{f_{cu,10}} + 1.438\right) \times 10^4$	4.308	4.448	4.582	4.710	4.834	4.953

8.4 SHRINKAGE OF SUPER-HIGH-STRENGTH HIGH PERFORMANCE CONCRETE

8.4.1 Shrinkage characteristics of super-high-strength high performance concrete

The autogenous shrinkage of concrete is a volume deformation that inevitably occurs during its setting and hardening. When the concrete is unconfined, the shrinkage does not lead to negative consequences, but in practice, the concrete is restricted due to the influence of the foundations, steel bars, or other neighboring influences, so the shrinkage of concrete is restricted and tensile stress is produced in concrete. When such tensile stress exceeds the tensile strength of concrete, the concrete will crack. Such shrinkage cracking that is not caused by loading will affect the bearing capacity of the structure as well as its safety and service life in construction. For example, the crack could cause the concrete to leak, and water, gas, and other aggressive substances can penetrate into the depth of the concrete through the cracks, resulting in corrosion of the steel reinforcement or corrosion of concrete as well as accelerating damage due to freeze–thawing.

The technical way to prepare super-high-strength high performance concrete is by addition of active mineral admixtures, increasing binder content, introduction of superplasticizers, and a significant decrease in water/binder ratio. Compared to traditional ordinary concrete, even with high strength high performance concrete, such measures lead to large differences in setting and hardening, and the internal structure of super-high-strength high performance concrete. Such differences give super-high-strength high performance concrete its own characteristics such as large plastic and autogenous shrinkage.

Plastic shrinkage occurs in the plastic stage before the concrete hardens, the shrinkage being due to the loss of water from the surface of the concrete. In the freshly mixed and compacted state, the space between the particles in concrete is filled with water. In the case of inadequate curing, when the rate of water loss from the surface exceeds the rate of water transfer from the inside of the concrete to the surface, a negative pressure is formed in the capillaries, causing the plastic shrinkage of concrete. In super-high-strength high performance concrete the water/binder ratio is low, which means the free water content is small, and in addition, the active mineral admixture usually has a high water demand so in practice it is easy for plastic shrinkage to occur leading to cracking. To prevent plastic shrinkage requires lowering the rate of water loss from the concrete. The measures usually applied include wind proofing, lowering the temperature, and retarding the setting time of the concrete. In order to keep the surface wet before final setting of the concrete, it is normal, for example, to cover the surface with plastic film, spray a curing agent on the surface, and ensure wetting continues after final setting.

As described above, drying shrinkage is caused by water loss from concrete due to its transfer to the outside surface. Autogenous shrinkage is caused by self-desiccation of the concrete, which occurs after casting and molding of concrete when no additional water is supplied so only the water introduced during mixing is available. Even if the loss of the existing water to the environment is prevented, the amount of water inside the concrete is reduced as hydration develops, reducing the relative humidity inside the concrete leading to a phenomenon called *self-desiccation* or *self-desiccation of concrete*. This self-desiccation can also cause unsaturation inside the capillaries where menisci form causing a capillary pressure difference, leading to self-shrinkage. In ordinary concrete with its high water/cement ratio, such shrinkage is small; for example, in concrete used in dam construction, the self-shrinkage for 1 month was only 40×10^{-6} and for 5 years 100×10^{-6},[10] which attracted little attention and in practice was only considered as part of the total shrinkage. With the development and application of high strength concrete, especially super-high-strength concrete, it has been found that their self-shrinkage is much more than that of ordinary concrete. Due to the low water/binder ratio and small water content, the amount of free water available for the hydration of cement is small. In addition, as the strength of high strength and super-high-strength concrete at early ages develops rapidly, much of the water is consumed; therefore, the self-shrinkage of high strength and super-high-strength concrete develops quickly at an early age.

Sellevold and Justnes[52] have studied the variation of relative moisture in cement paste during the process of its setting and hardening. Their study showed that the relative moisture content of concrete decreased with age and with a decrease of the water/binder ratio (Figure 8.11). Especially at an early age, the rate of reduction was greater, the cement paste with water/binder ratio of 0.20 showing a reduction down to 78% at 1 year, whereas for a water/binder ratio of 0.40 it was down to 87% at 1 year.

Jiang Zhen-Wu et al.[63] have studied the self-shrinkage development of the concrete with water-to-cement ratios in the range of 0.2 to 0.8, and obtained similar results; that is, with time, the lower the water/cement ratio, the faster the internal relative humidity decreased and the lower the internal relative humidity. Different mineral admixtures have different effects on the self-shrinkage development of concrete.

8.4.2 Shrinkage and self-shrinkage of super-high-strength high performance concrete

Wang Yongwei[54] studied the shrinkage and self-shrinkage deformation of super-high-strength high performance concrete. Considering the rapid

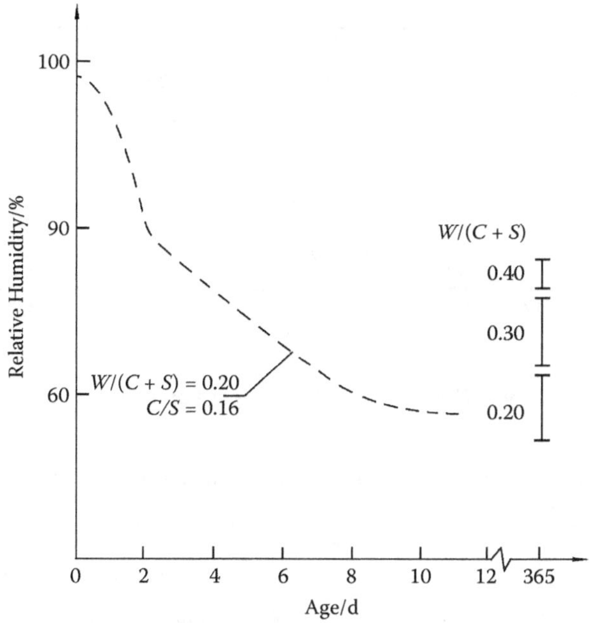

Figure 8.11 Variation of relative moisture of cement paste and relative moisture value after 1 year.

self-desiccation of super-high-strength concrete and its rather large shrinkage at an early age, it is unreasonable to apply the standard method used for ordinary concrete to determine its shrinkage. The standard method requires the specimen should be demolded 1 d after molding, and cured under standard conditions for a further 2 d, when the length of the specimen is measured and called the *initial length*. Thus, more than 2 d of shrinkage may be omitted and, in particular, the self-shrinkage deformation. Some researchers have suggested that the initial length of the specimen is preferably determined after initial set of concrete. There are certain problems with this realization and no standard method has been developed yet. Wang Yongwei[54] used the following method: the mold was removed when the strength of the concrete reached 3 to 5 MPa and the measuring head was glued onto the specimen and an initial length measured before it was moved to the shrinkage room and held at constant temperature (20 ± 3°C) and relative humidity (60 ± 5%). Dimensions of the specimens were 100 mm × 100 mm × 515 mm. Some of the specimens were cured in the standard curing room and some in water to observe the deformation change between these two different conditions. To determine the proper time to remove the molds, the development of strength at hourly intervals for different super-high-strength concrete mixes was tested before the shrinkage test as shown in Figure 8.12.

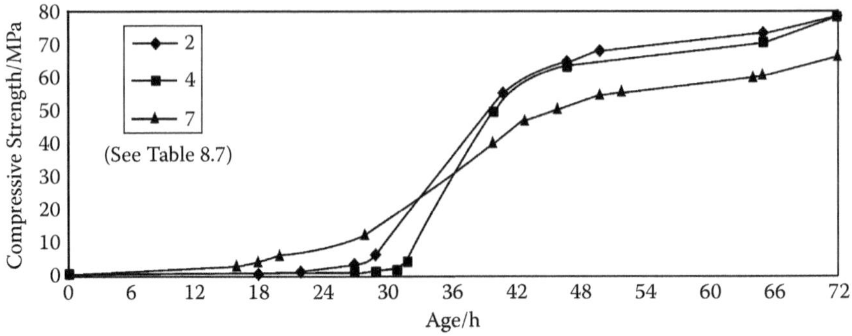

Figure 8.12 Development of hourly strength at early age of super-high-strength concrete.

Table 8.7 Mix proportion, fluidity, and strength of super-high-strength concrete for shrinkage test

No.	*W/B*	*Binder (kg/m³)*	*Binder composition (%)*			*SP (%)*	*Fluidity (mm)*		*Compressive strength (MPa)*		
			Cement	*SF*	*UEA*		*Slump*	*Spread*	*MR*	*3d*	*28d*
1	0.20	650	90	10	0	40	105		5	89.2	123.6
2	0.20	650	78	10	12	40	20		5.4	79.2	114.6
3	0.25	650	90	10	0	40	270	795	7	78.3	122.2
4	0.25	650	78	10	12	40	260	735	5.1	78.5	114.6
5	0.25	650	76	10	14	40	260	710	5.1	67.7	112.3
6	0.30	650	90	10	0	40	260	590	3.4	69.7	108.8
7	0.30	650	78	10	12	40	260	485	3.8	66.7	101.4

Note: W/B = water/binder ratio; SF = silica fume; SP = sand percentage; MR = after removing the mold.

The mix proportions of super-high-strength high performance concrete used for the shrinkage test are given in Table 8.7. The total binder content was 650 kg/m³, with 10% silica fume, and water/binder ratios of 0.2, 0.25, and 0.3. To compensate for the shrinkage of super-high-strength concrete, a further four groups of specimens were prepared containing UEA expansion agent. The materials used were Chongqing 525 OPC, Guiyang Qinzhen silica fume, UEA-H from Chongqing Jiangbei Special Building Materials Factory, medium sand with a fineness modulus of 2.8, crushed limestone (maximum size 20 mm), and a naphthalene sulfonate superplasticizer.

It can be seen from Figure 8.12 that for a water/binder ratio of 0.2 and with addition of UEA, the hourly strength developed slowly, reaching strength of 3 to 5 MPa after about 26 to 28 hours. For mix No. 4, in which the water/binder ratio was 0.25 and with UEA, 30 to 32 hours were needed. For the mix with water/binder ratio of 0.3 and UEA, only 16 to

20 hours were required while only 12 to 18 hours were required for the mix without UEA.

As the specimen in the mold is supported by the bottom and side plates, the intensive shrinkage of concrete is restricted, so that during this period, only the first part of setting and hardening is observed. In Figure 8.12, it can be seen that the hydration occurs slowly and the strength development is slow. The specimen is covered by a film and there is plenty of water in the concrete during this period, so that self-desiccation is low and the autogenous shrinkage small. Therefore, it is reasonable to take this time for measuring the initial length of the specimen.

Except for the total shrinkage of concrete, Wang Yongwei[54] has measured self-shrinkage of mixes Nos. 3, 4, and 5. The results are given in Table 8.8 and Figures 8.13 and 8.14 from where the following can be seen:

1. The total shrinkage of super-high-strength high performance concrete is large. The shrinkage of concrete without UEA reached (1229 to 1579) $\times 10^{-6}$ while with UEA it was slightly less.
2. The shrinkage of super-high-strength high performance concrete occurs mostly at early ages, with the shrinkage at 3 d more than half (71.6%) that recorded at 180 d but the rate of increase reduced with time so at 56 d it was only 54.5%.
3. To compare the total shrinkage of three mixes (mixes Nos. 1, 3, and 6) with water/binder (w/b) ratios of 0.20, 0.25, and 0.30, mix No. 3 with a w/b ratio of 0.25 had the maximum shrinkage (-1579×10^{-6}) followed by mix No. 6 with a w/b ratio of 0.30 (-1305×10^{-6}), while the minimum was that of mix No. 1 with a w/b ratio of 0.20 (-1229×10^{-6}). The order of the test results does not coincide with that of the w/b ratio and shows that there are multiple factors influencing the shrinkage, although the shrinkage due to self-desiccation increased with a reduction of w/b ratio.[10,64] If self-desiccation were the only factor, the total shrinkage of concrete with w/b ratio of 0.20 would be expected to be more than that with w/b ratio of 0.25. However, autogenous shrinkage is only a part of total shrinkage (although an important part). First, concrete with lower water content looses less water to the environment due to drying, so the drying shrinkage is less. As pointed out by Bian Rongbing,[65] capillary pores of 2.5 to 50 nm are full with water. The loss of water inside the capillary, whether due to internal or external drying, lowers the water surface enlarging the curvature of the meniscus and producing capillary pressure within the pore due to the surface tension of the water, leading to shrinkage of the concrete. When the capillary pore is more than 50 nm, the capillary tension can be neglected, while for capillaries with diameters less than 2.5 nm, no meniscus is formed.

Table 8.8 Shrinkage and self-shrinkage of super-high-strength concrete

No.	W/B	UEA	T or S	Shrinkage at the age of ($\times 10^{-6}$) 3 d	7 d	14 d	28 d	56 d	90 d	180 d	S_{3d} S_{180d} (%)	$S_{180d} - S_{3d}$ ($\times 10^{-6}$)
1	0.20	0	T	–825	–922	–1008	–1115	–1169	–1200	–1229	67.1	–404
2	0.20	12	T	–715	–882	–940	–1002	–1045	–1103	–1172	61.9	–458
3	0.25	0	T	–1130	–1268	–1324	–1398	–1511	–1551	–1579	71.6	–449
			S	–538	–598	–612	–699	–777	–810	–821	65.5	–283
4	0.25	12	T	–1008	–1198	–1219	–1243	–1322	–1385	–1458	69.1	–450
			S	–497	–575	–605	–627	–612	–612	–650	76.5	–153
5	0.25	14	T	–1025	–1206	–1256	–1252	–1388	–1450	–1511	67.3	–486
			S	–497	–583	–621	–641	–622	–641	–724	68.6	–227
6	0.30	0	T	–817	–948	–1052	–1148	–1222	–1272	–1305	62.6	–488
7	0.30	12	T	–728	–959	–1055	–1125	–1216	–1297	–1355	54.5	–617

Note: W/B = water/binder ratio; T = total shrinkage; S = self-shrinkage; S_{180d} = shrinkage at 180 d; S_{3d} = shrinkage at 3 d.

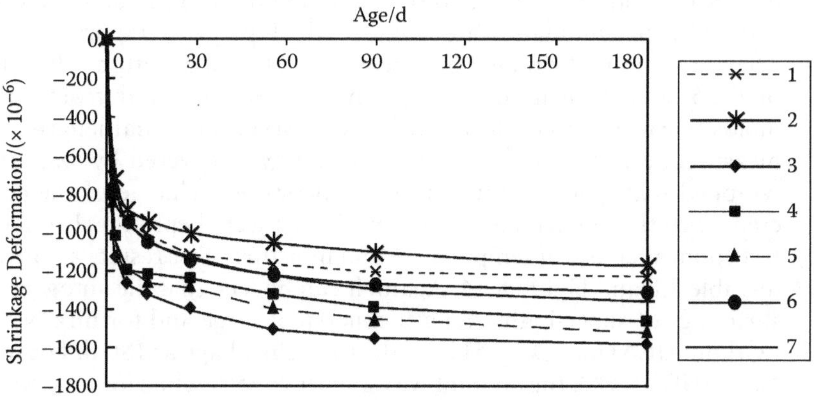

Figure 8.13 Shrinkage deformation of super-high-strength high performance concrete at different ages.

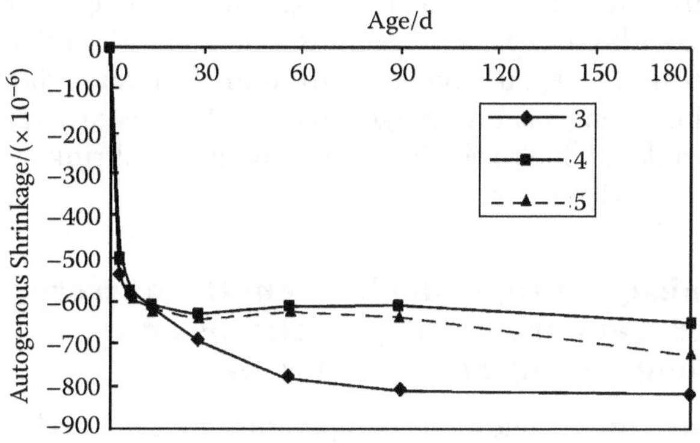

Figure 8.14 Autogenous shrinkage deformation of super-high-strength high performance concrete at different ages.

A study of the pore structure of super-high-strength high performance concrete (0.5 to 30 nm) shows that in mix No 1. (also W6 in Table 7.5) with a w/b ratio of 0.2, the volume of pores of diameter 0.5 to 3 nm is 12.27×10^{-3} ml/g, 37.55% of the total volume of 30.01×10^{-3} ml/g, while in mix No. 3 (also W7 in Table 7.5) with a w/b ratio of 0.25, the volume of pores of diameter 0.5 to 3 nm is only 8.63×10^{-3} ml/g, 28.78% of the total pore volume of 29.99×10^{-3} ml/g. As there are many pores without menisci in No. 1, it is understandable that as shown in Table 8.8, the shrinkage of mix No. 1 with a w/b ratio of 0.20 is less than that of mix No. 3. The total shrinkage of mix No. 6 (also W8 in Table 7.5) with a w/b ratio of 0.3 is less than that of mix No. 3 and may be due to the reduction of autogenous shrinkage caused by the abundant water content which plays an important role.

4. The autogenous shrinkage of mix Nos. 3, 4, and 5 with a w/b ratio of 0.25 was measured. The specimens were removed from their molds when the strength reached 3 to 5 MPa, the initial length was measured, and the surface of the concrete was covered by wax and wrapped with plastic film to prevent moisture exchange of the concrete with the environment. The specimens were kept in a shrinkage test room at constant temperature and moisture. Test results are given in Table 8.8 and Figure 8.14. Again, it can be seen that the autogenous shrinkage of super-high-strength concrete is large and for mix No. 3 (without UEA) makes up 51% of the total shrinkage at 180 d. For mix No. 4 (UEA 12%), the amount was smaller at 45% while for mix No. 5 (UEA 14%) it was 48%. In general, autogenous shrinkage makes up about 50% of the total shrinkage, much more than that for ordinary concrete where it is about 10% to 20%. Autogenous shrinkage developed rapidly in the first 3 d, making up about 65.5% to 68.6% of total shrinkage at 180 d. The autogenous shrinkage at early ages of concrete with an expansion agent is a little less than that of concrete without the agent, but compared to total shrinkage, the influence an expansion agent has on autogenous shrinkage is different to that on total shrinkage due to the fact that autogenous shrinkage of concrete develops slowly at later ages.

8.4.3 Shrinkage of super-high-strength concrete under standard curing conditions and swelling during curing under water

Under standard curing conditions (temperature 20 ± 3°C, relative humidity more than 90%), little of the water in super-high-strength concrete is lost to the environment. On the contrary, moisture from the atmosphere may condense onto the surface of super-high-strength concrete. Although the concrete is dense, it is full of an aqueous phase and water can be transferred into the concrete to compensate for the partial self-desiccation inside, greatly reducing the total shrinkage (Table 8.9 and Figure 8.15). Due to the difficulty in moving water to the outside, any shrinkage is mainly caused by self-desiccation and that is reduced significantly. For concrete containing 10% silica fume and a water/binder ratio of 0.25, the shrinkage at 180 d is only -421×10^{-6}, which is 26.7% of that of the control mix (-1579×10^{-6}) with the same mix proportions. The shrinkage at 3 d is rather large (-373×10^{-6}), making up 89% of that at 180 d but only 33% that of the control mix with the same mix proportions at 3 d held in the shrinkage test room (-1130×10^{-6}). At 14 d, the shrinkage became stable. Consequently, after

Table 8.9 Shrinkage and swelling of super-high-strength concrete under standard and water curing conditions

No.	W/B	UEA (%)	Curing condition	Deformation at the different ages ($\times 10^{-6}$) 3 d	7 d	14 d	28 d	56 d	90 d	180 d
3	0.25	0	Standard	−373	−410	−421	−414	−412	−408	−421
			In water	+99	+76	+89	+45	+87	+76	+122
4	0.25	12	Standard	−350	−365	−365	−332	−293	−282	−293
			In water	+227	+233	+289	+250	+334	+336	+403
5	0.25	14	Standard	−332	−350	−340	−332	−266	−235	−192
			In water	+240	+285	+299	+291	+334	+330	+452

Note: The numbering is consistent with that in Tables 8.7 and 8.8.

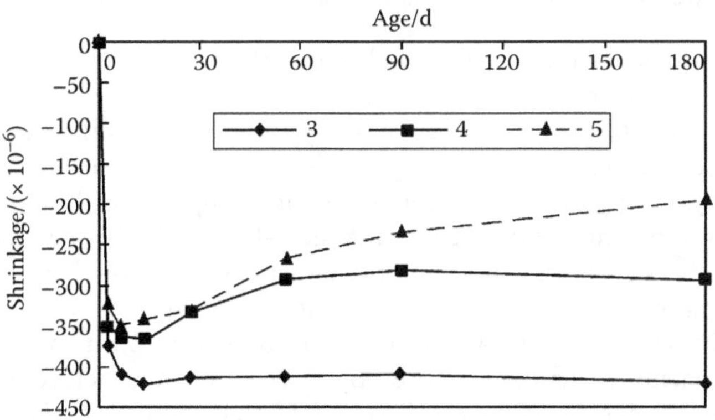

Figure 8.15 Shrinkage of super-high-strength concrete under standard curing conditions.

only 3 d of standard moist curing, the shrinkage of super-high-strength concrete is greatly reduced.

Concrete with addition of an expansion agent behaves differently under standard curing conditions. Its shrinkage reaches a maximum at 7 d, then decreases due to the expansion caused by the expansion agent. The reduction is greater when the addition increases from 12% to 14%. Total shrinkage at 180 d is about 46% of that of the control concrete.

Like ordinary concrete, when super-high-strength concrete is cured in water (in the water tank in a standard curing room) it swells (Table 8.9 and Figure 8.16), as the structure of super-high-strength concrete is dense and there will be a lack of water inside. Considerable water is consumed by hydration of cement as well as self-desiccation, which seemingly would lead to a large autogenous shrinkage. However, even the control concrete without an expansion agent (binder content of 650 kg/m^3, water/binder ratio

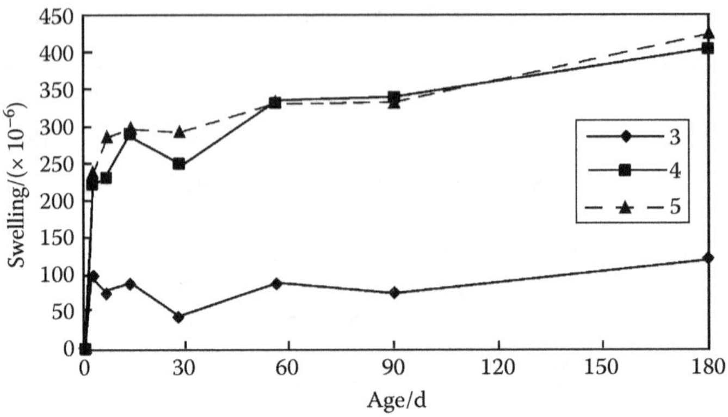

Figure 8.16 Swelling of super-high-strength concrete in water curing.

of 0.25, and silica fume of 10% replacement) undergoes large amounts of swelling at the age of 3 d (99×10^{-6}).

This swelling then slows after 28 d, so by 180 d, the swelling value is 122×10^{-6}. Consequently, water curing at an early age may reduce the shrinkage significantly, taking mix No. 3 as an example, compared with dry curing for 3 d, not only the large early age shrinkage (-1130×10^{-6}) (see Table 8.8) is eliminated but also a small amount of swelling develops, which is as expected.

The expansion of concrete containing an expansive agent increases remarkably over the first 14 d and and decreases by 28 d, which may be due to autogenous shrinkage caused by internal drying, before it increases again slowly. It can be seen from Figure 8.16 that after 28 d, the expansion of concrete containing an expansive agent is similar to that of control concrete without one, illustrating that during this period the expansive action is water swelling. The expansion of concrete with an expansive agent at 180 d reached $(403 \text{ to } 425) \times 10^{-6}$. This study showed that even in super-high-strength concrete with a low water/binder ratio, there will be marked expansion when a UEA expansive agent is used and the sample water cured. Such an agent can serve as a shrinkage compensator for super-high-strength concrete.

8.5 COMPENSATION FOR SHRINKAGE OF SUPER-HIGH-STRENGTH HIGH PERFORMANCE CONCRETE

As described in the previous section, because the dense structure of super-high-strength concrete makes it difficult for water to penetrate to the inside of super-high-strength concrete, the early strength that develops rapidly forms an obstacle to expansion, so there is little coordinated development of expansion with that of strength. In addition, silica fume added into the concrete consumes part of the $Ca(OH)_2$ necessary for forming the expansive ettringite. However, this is only another troublesome factor in producing super-high-strength shrinkage-compensated concrete.

Using a steel bar frame of ϕ10 to 100 mm × 100 mm × 356 mm to restrict shrinkage, a freely deformable specimen of 100 mm × 100 mm × 515 mm was prepared. The mix proportions and raw materials were the same as given in Table 8.7. The specimen was kept indoors after forming and covered with a plastic film until the strength of the concrete reached 3 to 5 MPa, when the mold was removed and the initial length of the specimen measured. After that, the specimen was soaked in water at 20 ± 3°C for 14 d, then cured in a shrinkage room under constant temperature 20 ± 3°C and relative humidity 60 ± 5%, before being measured at 3, 7, 14, 28, 90, and 180 d after the mold had been removed. The results are listed in Table 8.10 and Figure 8.17 from which it can be seen that

1. Mix Nos. 1, 3, and 6 show there may be certain restricted swelling in super-high-strength concrete that does not contain an expansive agent cured in water. Such swelling has little relationship to the water/binder ratio and the swelling rate is about 130×10^{-6}.
2. From the restricted deformation shown by mix Nos. 2, 4, 5, and 7, which contain an expansive agent, expansion is larger than that of control concrete. Taking 14 d as an example, mix No. 2 expanded

Table 8.10 Restricted and free deformations of super-high-strength concrete

No.	W/B	UEA	Deformation	Age Deformation ($\times 10^{-6}$)							Drop ($\times 10^{-6}$)
				3 d	7 d	14 d	28 d	56 d	90 d	180 d	
1	0.20	0	Restricted	127	130	107	−17	−63	−103	−137	−267
			Free	45	21	0	−186	−243	−305	−353	−398
2	0.20	12	Restricted	130	167	157	50	−7	−47	−137	−304
			Free	115	126	142	−64	−136	−210	−311	−453
3	0.25	0	Restricted	100	127	73	−10	−83	−123	−153	−280
			Free	99	76	89	−130	−264	−320	−355	−454
4	0.25	12	Restricted	170	217	227	153	40	23	−107	−334
			Free	227	233	289	140	−8	−111	−231	−520
5	0.25	14	Restricted	180	210	257	143	60	3	−93	−350
			Free	240	285	299	137	−18	−121	−264	−563
6	0.30	0	Restricted	133	130	113	3	−90	−123	−157	−290
			Free	78	60	14	−153	−295	−353	−392	−470
7	0.30	12	Restricted	270	383	297	210	107	27	−87	−470
			Free	363	390	361	206	50	−47	−173	−563

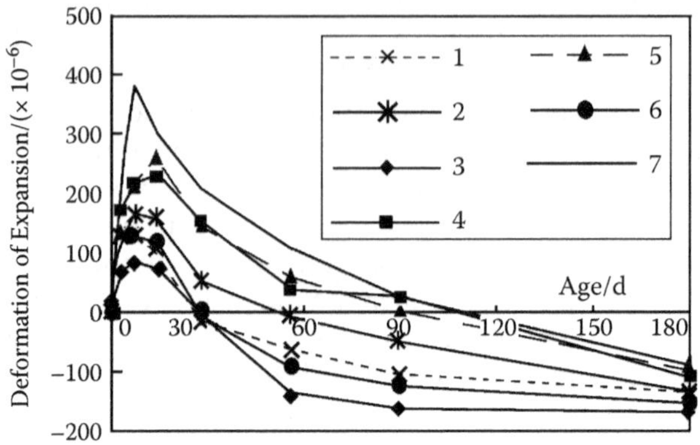

Figure 8.17 Restricted deformation of super-high-strength high performance concrete.

47% more than mix No. 1, mix No. 4 expanded 211% compared to mix No. 3, mix No. 5 expanded 252% more than mix No. 3, and mix No. 7 expanded 163% compared to mix No. 6. This large expansion at early ages is beneficial to compensate for the problems of large shrinkage of super-high-strength concrete at early ages.

3. When mix Nos. 2, 4, and 7 are compared with mixes with the same formulation but an increased water/binder ratio, the restricted expansion increased. Using the 14 d figures, mix No. 4 expanded 45% more than mix No. 2; mix No. 7 more than 31% that of mix No. 4 and more than 89% that of mix No. 2. Consequently, the lower the water/binder ratio, the faster the development of concrete strength at early ages and the higher the strength, the expansion proceeds with more difficulty.
4. When mix No. 5 is compared with mix No. 4, expansion with 14% of addition is only slightly larger than that with 12% addition.
5. Shrinkage-compensated concrete is one with an expansion at 14 d of more than 150×10^{-6} measured according to the China specification, and a shrinkage at 180 d of less than 400×10^{-6}. The restricted deformations of mix Nos. 2, 4, 5, and 7 show that at 14 d, all samples of super-high-strength concrete with an expansive agent were more than 150×10^{-6} and at 180 d the shrinkages were less than 400×10^{-6}. Therefore, by adding 12% to 14% of UEA expansive agent, shrinkage-compensated, super-high-strength concrete can be obtained.
6. Muguruma[66] considered that it was not necessary to have a large expansion; what was more important was the small drop after expansion. In such a case, the resistance to cracking will be high. The calculated shrinkage drop of super-high-strength concrete (shrinkage drop =

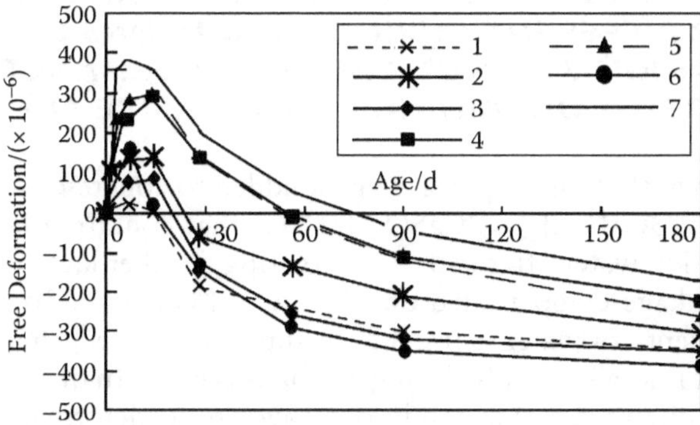

Figure 8.18 Free deformation of super-high-strength high performance concrete.

maximum shrinkage in 180 d – maximum expansion in 180 d) is given in Table 8.10, which shows that with an increase in water/binder ratio, the shrinkage drop increased. Therefore, this should be considered in the design and application of super-high-strength shrinkage-compensated concrete.

7. In Table 8.10 and Figure 8.18, the free deformations of super-high-strength concretes water cured for 14 d then kept in a controlled room are given. It can be seen from Figure 8.18 that the curves for free deformation of super-high-strength concrete coincide with that for restricted deformation, the only difference being their values.

When compared with restricted deformation, the free shrinkage of super-high-strength concrete is higher with the total shrinkage drop also larger. The free expansion values are more uncertain and may be larger or smaller. In case of concrete with a water/binder ratio of 0.30 and 0.25 and containing an expansive agent, free expansion is larger than the restricted one, understandable due to the difficulty of restricted expansion. However, in the absence of an expansive agent, free expansion is less than that of a restricted sample. In the case of concrete with a water/binder ratio of 0.2 and in the presence of an expansive agent, free expansion is still less than that of a restricted sample, and in the absence of an expansive agent, free expansion is even less. The cause of such a phenomenon should be investigated. Probably, part of the shrinkage is produced before measurement of the initial length (waiting until the concrete strength reaches 3 to 5 MPa), which imposes an elastic compression on the steel bar: this is recovered during water soaking and assists the expansion of the specimen with restricted deformation.

8.6 DEFORMATION AND STRENGTH DEVELOPMENT UNDER DIFFERENT CURING CONDITIONS OF SUPER-HIGH-STRENGTH HIGH PERFORMANCE CONCRETE CONTAINING AN EXPANSIVE AGENT

It is well known that one expansive agent added to concrete will form ettringite crystals ($3CaO{\cdot}Al_2O_3{\cdot}3CaSO_4{\cdot}32H_2O$) that contain a large amount of crystallization water after hydration creating a rather large expansion to produce shrinkage-compensated or self-stressed concrete. To form such a compound, plenty of water is needed. In super-high-strength concrete, the water content is low so supply during curing is crucial. In order to ascertain the influence different curing conditions have on the deformation of super-high-strength high performance concrete, concrete was tested, in which the binder content was 650 kg/m^3, silica fume 10%, water/binder ratio 0.25, and 12% expansive agent was added (Table 8.7, mix No. 4). The development of deformation under different curing conditions was followed. Conditions were drying under constant temperature and moisture, drying with constant temperature after 14 d of water curing, hermetic conditions at constant temperature and isolation from moisture, standard curing, and water curing. Test results are given in Figure 8.19 where it can be seen that the availability of water affects the expansion of super-high-strength shrinkage-compensated concrete. The expansion can only be fully realized under total water curing. Under drying or standard moisture conditions, there will be shrinkage. At 180 d the deformation reached -1458×10^{-6}

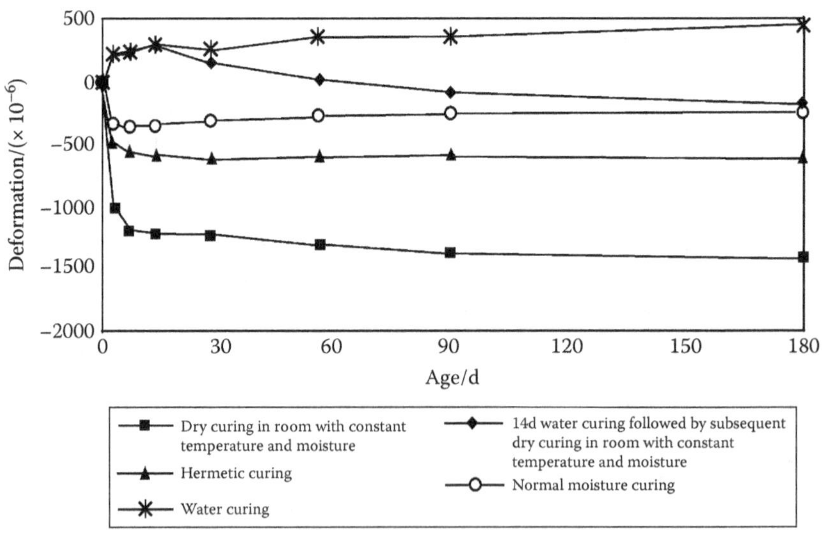

Figure 8.19 Development of deformation of super-high-strength concrete with expansive agent under different curing conditions.

(drying environment), -65×10^{-6} (moisture isolation environment), -293×10^{-6} (standard moisture environment), -231×10^{-6} (curing under constant temperature and moisture after 14 d of water curing), and 403×10^{-6} (water curing). Therefore, it is important to ensure early water curing for super-high-strength concrete containing an expansive agent.

After addition of an expansive agent, a great deal of ettringite is formed in the concrete. Whether the ettringite is stable, or whether the expansion process is coordinated with strength growth, will affect the development of concrete strength. Consequently, the compressive strength at different ages for super-high-strength concretes with or without expansive agent was measured, and the results are given in Table 8.11 and Figures 8.20 and 8.21.

Table 8.11 Strength development of super-high-strength concretes

No.	*W/B*	*UEA replacement (%)*	*Curing*	*Compressive strength at the age of (MPa)*						
				3 d	*7 d*	*14 d*	*28 d*	*56 d*	*90 d*	*180 d*
1	0.20	0	Standard	89.2	104.3	113.0	123.6	139.3	141.2	142.9
2	0.20	12	Standard	79.2	101.0	108.4	114.7	135.3	136.4	137.8
3	0.25	0	Standard	78.3	103.4	110.4	122.2	126.4	131.4	132.4
			[a]	77.3	94.8	110.3	128.0	137.2	144.9	145.6
4	0.25	12	Standard	78.3	99.4	108.0	114.6	123.6	125.7	127.4
			[a]	78.4	97.4	104.6	125.8	130.7	142.8	143.1
5	0.25	14	Standard	67.7	94.4	107.7	112.3	117.1	121.4	121.6
			[a]	65.8	87.8	100.3	113.0	124.5	135.9	137.4
6	0.30	0	Standard	69.7	87.8	97.4	108.8	111.7	116.2	118.2
7	0.30	12	Standard	66.7	84.1	85.4	101.4	106.3	110.6	112.2

[a] Curing under constant temperature and moisture after 14 d of water curing.

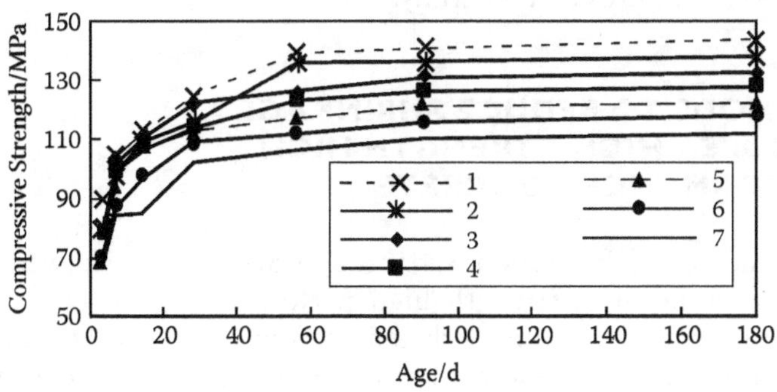

Figure 8.20 Strength development of super-high-strength high performance concrete at different ages under standard curing conditions.

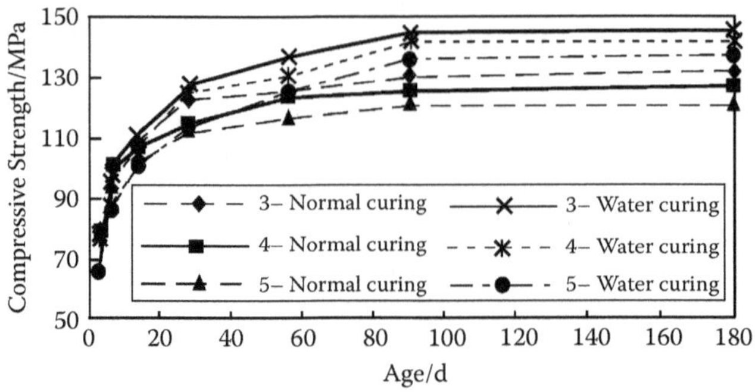

Figure 8.21 Development of strength of super-high-strength high performance concrete under different curing conditions.

It can be seen from Table 8.11 and Figure 8.20 that for super-high-strength concrete without an expansive agent, the strength developed steadily, whereas for super-high-strength concrete with an expansive agent, the growth of strength in the first 56 d is variable, maybe due to the influence of an expansion agent but did not decrease. After 56 d the strength developed steadily. At any age, the strength of concrete containing an expansive agent was lower than that without one. With increasing dosage, the drop of strength was greater, maybe due to the decrease of cement content in the binder, but in general, the influence of an expansive agent on the strength of super-high-strength concrete was not significant.

It can be seen from Table 8.11 and Figure 8.21 that during 14 d of water curing, the strength of concrete is slightly less than that under standard curing. However, after water curing and continued curing under drying conditions in the shrinkage room at constant temperature and humidity, the strengths of concrete both with and without an expansion agent were apparently higher than those held under standard curing conditions. The reason for that requires further study.

8.7 METHODS TO REDUCE SHRINKAGE OF SUPER-HIGH-STRENGTH HIGH PERFORMANCE CONCRETE

The study described in Section 8.4 through Section 8.6 showed that the shrinkage of super-high-strength high performance concrete was large, especially in the first 3 d, and reached 54.5% to 71.6% of the shrinkage at 180 d, but it had no apparent relationship with the water/binder ratio. The shrinkage at later ages (3 to 180 d) was also not significant and consisted of about 500×10^{-6}. In the shrinkage of super-high-strength concrete, there is a large percent of autogenous shrinkage; at water/binder ratio of 0.25, the autogenous shrinkage is about 50%.

Though the shrinkage of super-high-strength concrete is relatively large, it does not mean there is no way to solve it. It is important to offer early and plentiful water. Because whatever the type of concrete, there will be swelling with water curing, and there is no exception for super-high-strength high performance concrete (Figure 8.16), although there is high compactness of super-high-strength high performance concrete with very low permeability, because with curing in water, the water still can penetrate into the concrete causing it to deform by swelling.

Swelling deformation is due to an expansive pressure, caused by the buildup of a water film on the sub-microcrystals as a result of water adsorption on the gel formed by hydration of the cement. This pressure separates the gel particles by overcoming the cohesive force. With a continuous supply of water, entering water not only occupies the space caused by the contraction due to the hydration of cement (volume of the hydration product < hydrated clinker + the water consumed, reduction of volume per 100 g of cement is 8 to 10 ml), making autogenous shrinkage caused by self-desiccation during hydration of concrete impossible and no drying shrinkage.

Therefore, the autogenous shrinkage and drying shrinkage of super-high-strength high performance concrete can be overcome by plenty of water giving rise to swelling of concrete and preventing a worsening of durability as a result of cracking due to the shrinkage of concrete. Here, two rules should be followed: (1) curing should be started as early as possible after casting and molding and particularly after initial set, and (2) the water supply should be plentiful. This is because initially there is plenty of water in the concrete, and further water allows a gradual penetration to the inside of the concrete. If the supply of water starts after the appearance of a gaseous phase has appeared, then the aqueous phase becomes discontinuous, and water penetration becomes a penetration–diffuse process and water movement is difficult. To ensure a continuous water supply, moist or water curing can be applied. Moist curing uses a water-absorbable material such as gunny sack or hessian immersed into water until saturated, covering the surface of the concrete (two layers) and continuously wetted at intervals to remain water saturated during curing; water curing means that the concrete is immersed in water and stored this way. Gan Changchen et al.[67] studied the application of both these methods in practice and achieved excellent results. C20–C40 concrete with high water permeability was enhanced to class S30–S39 and more. This demonstrates that moist and water curing are effective for overcoming the shrinkage of ordinary concrete, though there have been no reports of the effectiveness for super-high-strength high performance concrete. According to the laboratory experimental data of Wang Yongwei[54] given in Table 8.9, the author believes that early application of moist curing or water curing enables a large amount of the autogenous shrinkage and drying shrinkage of super-high-strength concrete to be reduced, effectively preventing shrinkage cracking of super-high-strength high performance concrete.

As described above, even 3 d of water curing can diminish to a great extent the shrinkage of super-high-strength high performance concrete at early ages. To ensure super-high-strength high performance concrete does not crack, moist or water curing should preferably be no less than 14 d.

Addition of an expansive agent can produce super-high-strength shrinkage-compensated concrete, as well as solving the problem of large shrinkage in super-high-strength high performance concrete.

Modern shrinkage reducing agents offer another intensive and effective technical way to diminish the shrinkage of super-high-strength high performance concrete. Many researchers have shown[65,68] that addition of a shrinkage-reducing agent can reduce shrinkage up to 40% to 50% and more. If an expansive agent is used in combination with a shrinkage-reducing agent, the effect of shrinkage reduction will be better.[69]

8.8 CREEP OF SUPER-HIGH-STRENGTH HIGH PERFORMANCE CONCRETE

The author has not investigated the creep of super-high-strength concrete due to lack of funding and equipment. There has been little in the way of study on this subject elsewhere in the world, so the data are not systematic and well established. In Japan, Komuro and Kuroiwa et al.[70] have carried out the first experiments and a brief introduction is given. The mix proportions of concrete they used are given in Table 8.12. The curves for strength development of the specimens with time are shown in Figure 8.22 for cores drilled from columns 800 mm × 800 mm of three mixes H20, N22, and N27. The measurement was carried out on 7.5-year-old specimens but even at this age the strength continued to increase.

Creep testing under compression was carried out according to JIS on three cylinder specimens with a diameter of 100 mm and a height of 200 mm. The load was applied by a jack, controlled by an automatic device for oil pressure control (Figure 8.23). The ratio of load to strain (strain from vertical load divided by compressive strength at 28 d) was taken as 1/3 of long-term permitted strain of concrete. Figure 8.24 shows the relationship between the loading age and creep of 150-MPa super-high-strength concrete prepared by the researchers and based on the mix proportions given in Table 8.12. With an increase in age, the creep of 150-MPa concrete was smaller than that of 100-MPa concrete due to greater compaction of the former. The researchers studied the creep factor of concrete with a water/binder ratio of 0.3 to 0.6 and compressive strengths of 50.3 to 119.1 MPa (Figure 8.25).

Table 8.12 Mix proportion of super-high-strength concrete

No.	*Aggregate*	*Cement*	*W/B*	*Binder/ (kg/m³)*	*Binder composition (kg/m³)*			*Water content (kg/m³)*	*Stone content (kg/m³)*	*Sand content (kg/m³)*
					Cement	*Slag*	*SF*			
H15	Crushed andesite, crushed sand	Early strength PC	0.15	1000	700	200	100	150	871	461
H20	Crushed gneiss, crushed sand	Early strength PC	0.20	750	525	150	75	150	911	591
H24	Crushed gneiss, crushed sand	Early strength PC	0.24	626	480.2	125.2	62.6	150	918	694
N18	Crushed andesite, land sand	OPC	0.18	861	602.7	172.2	86.1	150	792	619
N20	Crushed hard sandstone	OPC	0.20	775	542.5	155	77.5	150	872	616
N22	Crushed hard sandstone	OPC	0.22	682	477.4	136.4	68.2	150	899	680
N27	Crushed hard sandstone	OPC	0.27	556	389.2	111.2	55.6	150	899	792

Note: W/B-water/binder ratio, SF - silica fume, the sand for mixes N20, N22, and N27 is crushed sandstone sand: land sand = 5:5.

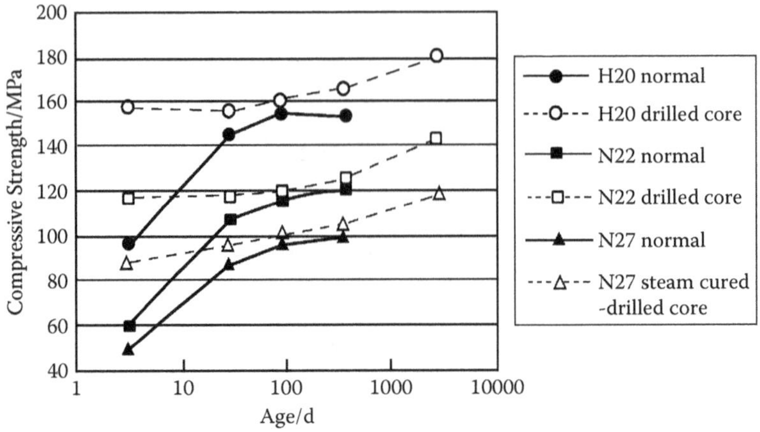

Figure 8.22 Relationship between compressive strength and age of super-high-strength concrete.

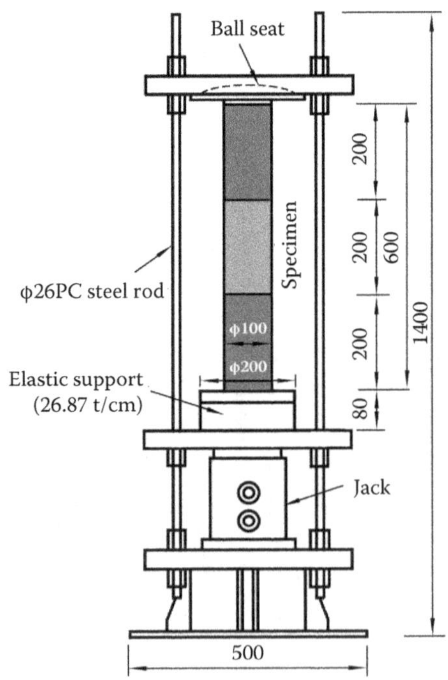

Figure 8.23 Creep test installation.

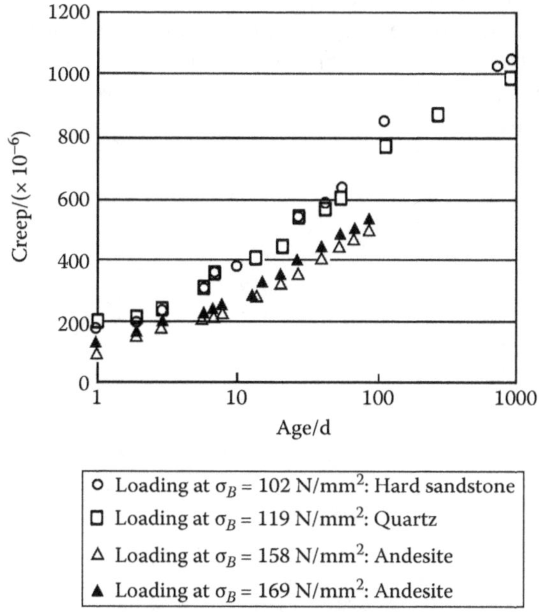

Figure 8.24 Relationship between loading age and creep of 150-MPa super-high-strength concrete.

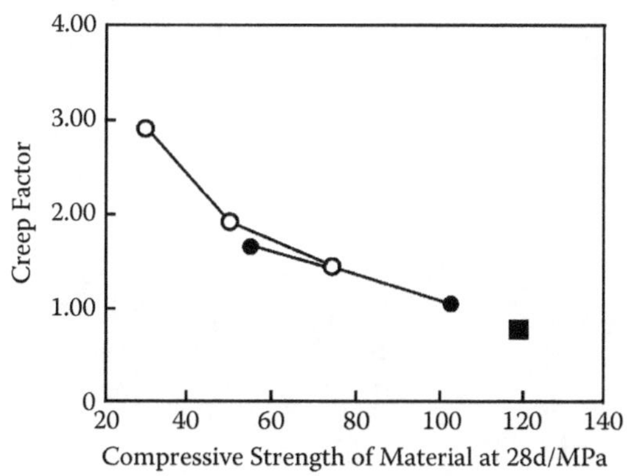

Figure 8.25 Relationship between creep factor and compressive strength.

With increasing compressive strength, the elastic deformation increased and the related creep factor decreased.

The creep behavior of concrete depends on its microstructure. When compared with ordinary concrete, high strength and super-high-strength concretes have smaller creep. When preparing high strength and super-high-strength concretes, the water/binder ratio is lower; this leads to a lower degree of hydration of cement and less hydration products. In addition, due to the self-desiccation that occurs, the free water content in concrete is even less. All these lead to a lowering of the creep of high strength and super-high-strength concretes.[71]

Chapter 9

Durability

9.1 PERMEABILITY OF SUPER-HIGH-STRENGTH HIGH PERFORMANCE CONCRETE

The durability of concrete is closely related to its permeability. Concrete with high permeability has more total porosity and pores with the greatest mean pore diameter. These are more open and interconnected so aggressive media easily penetrate into the concrete causing corrosion and damage, leading to reduced durability.

As described in Chapter 7, the water/binder ratio of super-high-strength high performance concrete is very low (0.2 to 0.3), so the pores are fine and the diameter of the mean pore is very small; overall, there is a low total porosity. Therefore, the permeability of super-high-strength high performance concrete is expected to be very low.

Powers[42] studied the permeability of hardened cement pastes with different water/cement ratios (Figure 9.1). When the water/cement ratio was reduced from 0.7 down to 0.6, the permeability coefficient dropped rapidly. When the ratio was reduced from 0.6 down to 0.5, the permeability coefficient was lowered to less than 10×10^{-14} m/s, while reducing it from 0.5 to 0.4 gave very low permeability, and when the water/cement ratio was lowered to 0.3, the permeability was close to zero.

As shown in Figure 9.1, the permeability of concrete determined by the accelerated chloride test follows the same pattern, that is, the lower the water/cement ratio, the lower the permeability of the chloride ion, especially when silica fume is added at a water/binder ratio less than 0.3, when the permeability of the chloride ion is close to zero.

Nagataki[9] studied the air permeability of concrete and showed that air permeability correlated with that of water penetration (Figure 9.2). While a decrease of the water/binder ratio lowered the air permeability, it was higher than water permeability by 1 to 2 orders of magnitude. Addition of silica fume in combination with a superplasticizer gave the best effect.

The author and his colleagues investigated the permeability of high strength and super-high-strength concretes using the method given in the

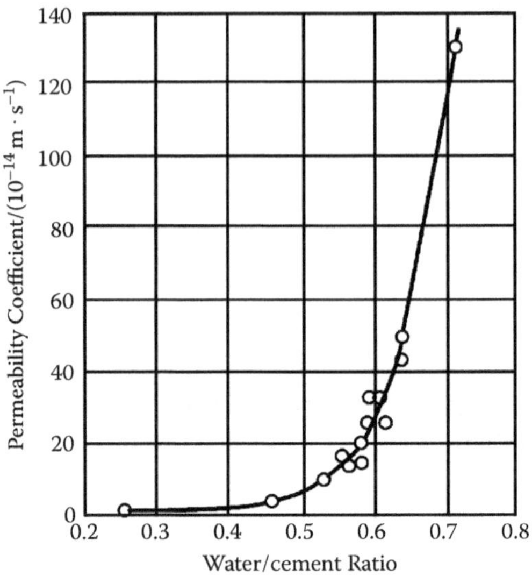

Figure 9.1 Permeation of hardened cement paste.

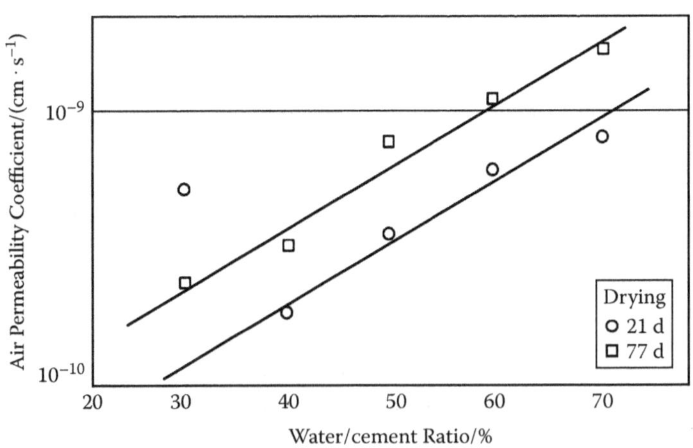

Figure 9.2 Air permeability of concrete.

Table 9.1 Impermeability of super-high-strength concrete

No.	Binder content (kg/m³)	s/(C+S) (%)	W/B	Admixture (L·m⁻³)	Slump (mm)	Spread (mm)	28 d R_{com} (MPa)	Impermeability pressure (MPa)	Height of penetration (mm)
1	600	10	0.20	21	160	—	103.9	4	0
2	600	10	0.25	21	260	640	102.9	4	0
3	600	10	0.30	21	270	770	74.9	4	2

Note: The maximum impermeability pressure of installation is 4 MPa.

Standard GBJ 82-85; test results, presented in Table 9.1, show that the impermeability of super-high-strength concrete is excellent. They also indicated that differences in the permeability of super-high-strength concrete could not be distinguished by the method given in GBJ 82-85.

9.2 FROST RESISTANCE OF SUPER-HIGH-STRENGTH HIGH PERFORMANCE CONCRETE

High strength and super-high-strength high performance concretes are prepared with a low water/binder ratio and the addition of an active mineral admixture. Their porosity is very low and their pore structure is excellent so their frost resistance is expected to be good. The literature data showed[50] that for concrete with a water/cement ratio of 0.35 and addition of an air-entraining agent, moist cured for 14 d with 50% rh curing for a further 76 d gave frost resistance of 7000 freeze–thaw cycles. The frost resistance of concrete without an air-entraining agent but with a water/cement ratio of 0.35 and cured in water for 28 d reached 3000 freeze–thaw cycles.[56]

Cao Jianguo and Li Jinyu et al.[28] studied the frost resistance of high strength concretes C60, C60 (air-entrained), and C80 as well as super-high-strength concrete C100. The mix proportions, fluidity, and concrete strength are given in Table 9.2. In that study, an accelerated test method was used and the variation in the specimens' frost resistance was expressed as a decrease of relative dynamic modulus of elasticity and mass loss of the specimen. The test results are given in Table 9.3, and Figures 9.3 and 9.4. It can be seen that C60 concrete did not have high frost resistance and did not meet the requirements of F300 but that a C60 concrete with an air-entraining agent can reach F1200. C80 high strength concrete and C100 super-high-strength concrete have excellent frost resistance and their frost resistance level can reach the requirements of F1200 and greater. As we have found, except for C60 where the concrete did not reach its limit of frost resistance, all other concretes met 1200 freeze and thaw cycles and

Table 9.2 Mix proportion, fluidity, and strength of high strength and super-high-strength concretes for the frost resistance test

Class of concrete	525 cement (kg/m³)	Water content (kg/m³)	W/B	Silica fume (kg/m³)	Slag powder (kg/m³)	Admixture		Air content (%)	Slump (cm)	R_{28c} (MPa)	R_{28b} (MPa)
						AEA 10^{-4}	WRA/%				
C60	500	130	0.26				1.4	1.5	9.0	65.9	10.56
C60 (AE)	500	130	0.26			0.2	1.4	4.3	10.0	57.2	10.46
C80	500	130	0.26	40	50		1.8	0.7	10.0	88.1	12.40
C100-1	500	130	0.22	50			1.6	0.7	10.0	105.2	15.6
C100-2	500	130	0.22	50			1.8	0.7	10.0	101.4	15.3

Note: W/B = water/binder ratio; AEA = air-entraining agent; WRA = water-reducing agent; R_{28c} = compressive strength at 28 d; R_{28b} = bending strength at 28 d.

Table 9.3 Experimental results for frost resistance of high and super-high-strength concretes

Grade	Index	Number of freeze/thaw cycles												
		25	*50*	*100*	*125*	*150*	*200*	*250*	*300*	*400*	*600*	*800*	*1000*	*1200*
C60	Relative dynamic E (%)	98.3	98.3	98.3	98.3	97.9	96.2	96.5	Cracking of specimen					
	Mass loss (%)	0	–0.1	–0.13	–0.13	–0.19	–0.19	–0.26						
	Compressive strength (MPa)							(270)60.0						
	Bending strength (MPa)							(270)2.10						
C60 (AE)	Relative dynamic E (%)	97.2		97.7			97.4			97.3	97.0	96.5	96.6	92.6
	Mass loss/%	0		0			0.03			0.1	0.27	0.45	0.52	0.61
	Compressive strength (MPa)													51.0
	Bending strength (MPa)													7.02
C80	Relative dynamic E (%)	98.3		98.3			98.2			96.3	95.7	96.9	96.4	92.7
	Mass loss (%)	0		0			0			0.01	0.04	0.06	0.07	0.07
	Compressive strength (MPa)													79.8
	Bending strength (MPa)													6.89
C100–1	Relative dynamic E (%)	98.8		98.7			98.5			98.4	95.5	97.9	97.4	96.9
	Mass loss (%)	0		0			0			0	0	0	0	0
	Compressive strength (MPa)													91.8
	Bending strength (MPa)													12.0
C100–2	Relative dynamic E (%)	98.2		98.1			98.1			98.1	98.2	98.1	98.2	98.2
	Mass loss (%)	0		0			0			0	0	0	0	0
	Compressive strength (MPa)													76.3
	Bending strength (MPa)													8.87

Note: E = modulus of elasticity.

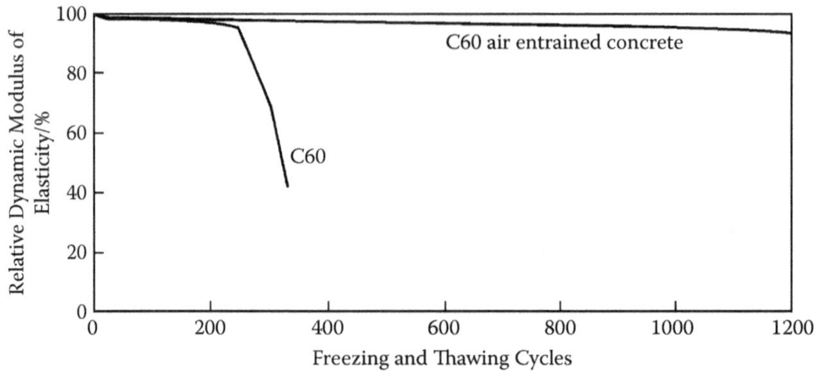

Figure 9.3 Relationship between reduction of relative dynamic modulus of elasticity and freeze/thaw cycles of C60 and C60 air-entrained concrete.

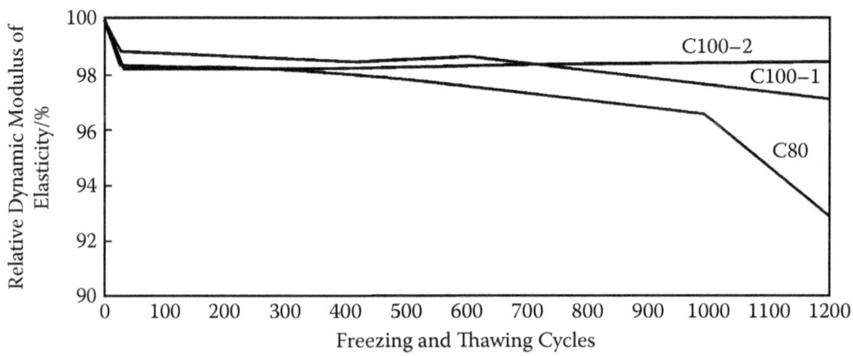

Figure 9.4 Relationship between reduction of relative dynamic modulus of elasticity and freeze/thaw cycles of C80 and C100 super-high-strength concretes.

the reduction of the relative dynamic modulus of elasticity and mass loss were both small.

Unlike the failure mode of ordinary concrete, the dynamic modulus of elasticity of C60 high strength concrete subjected to freeze–thaw cycling is only slightly reduced and there is no reduction in mass with even a slight increase. Failure is due to the formation and development of cracks in the concrete caused by fatigue stresses from the variations in temperature during the freeze and thaw cycles, the existence of these cracks leading to an increase in mass as water is absorbed, and the cracks cause a sudden brittle fracture. At this time, the relative dynamic modulus of elasticity decreases rapidly. The criterion for failure in ordinary and ordinary air-entrained concrete is a reduction of relative dynamic modulus of elasticity to 60% of the original and a mass loss of 5%. C60 air-entrained concrete has excellent frost resistance, as there are many fine air bubbles in the concrete that absorb the expansive pressure caused by water freezing and prevent the formation and development of cracks.

High strength and super-high-strength concretes greater than C80 have an unusual frost resistance exceeding 1200 freeze and thaw cycles, even without the addition of an air-entraining agent, while their relative dynamic modulus of elasticity is more than 90% and there is little or no mass loss.

The reduction in compressive strength of C60 concrete after 270 freeze and thaw cycles, C60 air-entrained concrete, C80 high strength concrete, and C100 super-high-strength concrete after 1200 cycles is not significant, but the reduction in bending strength is rather large, especially for the C60 non-air-entrained concrete. This index is more sensitive to freeze and thaw conditions compared to the relative dynamic modulus of elasticity, mass, and compressive strength loss.

Li Jinyu and Cao Jianguo[72] carried out continuous frost resistance tests on C100-2 super-high-strength concrete for up to 2165 freeze and thaw cycles, and found that the relative dynamic modulus of elasticity was reduced by 61.8% but the mass loss was still zero. In this case, apparent cracks appeared in the concrete, and the compressive strength was lowered to 62.0 MPa and the bending strength to 4.8 MPa.

9.3 RESISTANCE OF SUPER-HIGH-STRENGTH CONCRETE TO SULFATE CORROSION

The service conditions under which modern concrete is expected to perform are very severe, particularly when the concrete is used in marine engineering where there is often sulfate corrosion. Therefore, it is necessary to study the resistance of super-high-strength concrete to sulfate and seawater corrosion. Consequently, the author and his colleagues have carried out a preliminary study on this subject.[109]

The most corrosive sulfate is magnesium sulfate, because not only does sulfate corrosion occur but magnesium corrosion occurs as well. Typically, magnesium sulfate corrosion is studied using an aqueous solution of 2% $MgSO_4$. To study seawater corrosion, artificial seawater was prepared with a solution of NaCl (2.75%), $MgCl_2$ (0.39%), $MgSO_4$ (2.34%), $CaCl_2$ (0.05%), and $CaSO_4$ (0.11%). The dimensions of the concrete specimens were 100 mm × 100 mm × 100 mm. After the concrete was mixed and molded, the specimens were cured under standard conditions for 28 d, the strength determined, and then the specimens were soaked in an artificial seawater solution. The ratio of the specimen volume to that of the corrosive solution was 1:15 and the solution was changed every 28 d.

For comparison, a similar specimen was soaked in clean water and variations in strength variation measured at 180 and 432 d after soaking. The

Table 9.4 Test on resistance of super-high-strength high performance concrete to sulfate and seawater attacks

No.	Binder (kg/m^3)	Silica fume (%)	W/B	Fluidity (mm) Slump	Fluidity (mm) Spread	28 d Rcom (MPa)	Corrosive solution	Strength after corrosion for (MPa) 180 d	Strength after corrosion for (MPa) 423 d
1	600	10	0.20	90		103.9 (1.00)	Clear water	123.5 (1.19)	120.1 (1.16)
							$MgSO_4$	120.3 (1.16)	122.5 (1.18)
							Seawater	124.2 (1.20)	127.0 (1.22)
2	600	10	0.25	260	640	102.9 (1.00)	Clear water	102.1 (0.99)	109.4 (1.06)
							$MgSO_4$	117.9 (1.15)	116.2 (1.13)
							Seawater	111.1 (1.08)	117.6 (1.14)
3	600	10	0.30	270	720	75.1 (1.00)	Clear water	88.8 (1.18)	84.8 (1.13)
							$MgSO_4$	84.9 (1.13)	91.5 (1.22)
							Seawater	87.7 (1.17)	94.5 (1.26)

Note: In all mixes, a naphthalene sulphonate superplasticizer was added.

mix proportions, fluidity, 28 d strengths, and strengths after corrosion at different times are given in Table 9.4 and Figure 9.5.

It can be seen from Figure 9.5 that by soaking super-high-strength concrete in water or corrosive solutions, there was a general trend toward increased strength with the specimens soaked in $MgSO_4$ solution and seawater developing strength faster than those soaked in water, showing excellent resistance to sulfate and seawater attacks with no damage noted.

When the specimens were taken from the corrosive solutions, no edge or corner damage or dusting on the surface could be seen with the surface still smooth and neat, and the edges hard and strong. The reason why the strength of a corroded specimen was higher than that soaked in water needs further study. The author thinks that due to the addition of an active mineral admixture, any free $Ca(OH)_2$ in the hydration products of super-high-strength concrete was consumed, reducing the CaO/SiO_2 and CaO/Al_2O_3 ratios of calcium silicate hydrate and calcium aluminate hydrate.

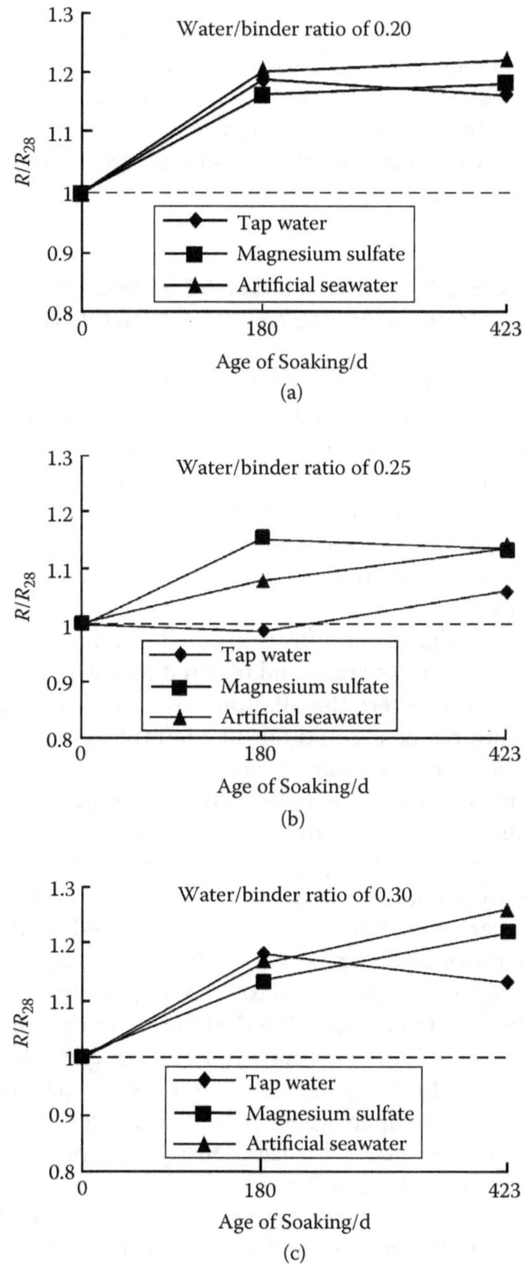

Figure 9.5 Sulfate resistance and resistance to seawater corrosion of super-high-strength high performance concrete.

When SO_4^{-2} or Mg^{+2} ions penetrate the concrete, the lack of $Ca(OH)_2$ or Al_2O_3, in addition to the low permeability of the concrete and little water in the concrete, makes it difficult to form sufficient deleterious materials such as gypsum ($CaSO_4 \cdot 2H_2O$), ettringite ($3CaO \cdot Al_2O_3 \cdot CaSO_4 \cdot 32H_2O$), or $Mg(OH)_2$. On the contrary, these reaction products have no deleterious effect but rather a beneficial effect, filling the pores and enhancing the strength of the concrete.

9.4 RESISTANCE OF SUPER-HIGH-STRENGTH HIGH PERFORMANCE CONCRETE TO CARBONATION

When exposed to the atmosphere, CO_2 penetrates through pores and microcracks into concrete and may react with $Ca(OH)_2$ or other hydrates to transform $Ca(OH)_2$ or C-S-H into $CaCO_3$ and amorphous SiO_2. Carbonation of $Ca(OH)_2$ leads to an increase in density and strength of the concrete, but the carbonation of calcium silicate hydrate has another result. Carbonation of high basic (high Ca/Si ratio) calcium silicate hydrate, tobermorite, and xonotlite leads to increased strength, while carbonation of CSH(I) leads to decreased strength.[19,73]

Carbonation has a significant effect on the durability of concrete. It will lower the alkalinity of the concrete, and may reduce the pH value from ≥12.5 down to less than 11.5, where the alkaline environment necessary for protection of steel reinforcement could not be guaranteed and the passive film of the reinforcement can be damaged leading to accelerated corrosion of the steel reinforcement. The volume expansion of the rust accelerates the damage of the concrete protective layer, and moisture and oxygen penetrate easily into the concrete, leading to further acceleration of the corrosion. When this deleterious cycle occurs, it leads finally to failure of the concrete structure.

The rate of concrete carbonation is usually related to the strength (or the water/binder ratio) of concrete. Figure 9.6 shows the depth of carbonation of concrete with strengths of 20 to 50 MPa, stored in the open air for 3 years.[74] It can be seen from Figure 9.6 that the lower the concrete strength, the greater the depth of carbonation. But for concrete with strength of 50 MPa, after 3 years of natural carbonation, the carbonation depth is zero.

For high strength and super-high-strength concretes, their water/binder ratio is lowered significantly with the density and permeability reduced due to the addition of a superfine mineral admixture, so even air hardly penetrates. Though a large part of $Ca(OH)_2$ will react with the additive introduced, there will be an increased amount formed due to the increased amount of cement in the high strength and super-high-strength concretes along with additional Na_2O and K_2O from some of the additives. The factors above enhance the carbonation resistance of high strength and super-high-strength concretes.

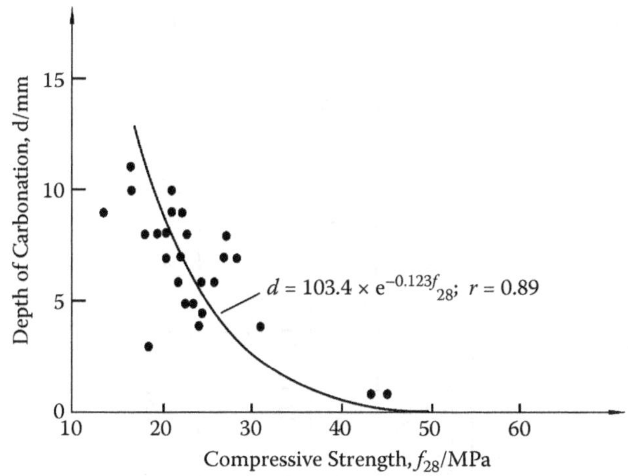

Figure 9.6 Carbonation depth as a function of compressive strength of concrete.

The author and his colleagues have carried out tests on the resistance of super-high-strength concretes to natural carbonation. The specimen was prepared and cured for 28 d under standard conditions before being placed on the roof of the laboratory (without covering) and allowed to carbonate naturally. The mix proportions of super-high-strength concrete samples and the results from natural carbonation for 1, 3, and 5 years are given in Table 9.5. It can be seen from the table that the depth of carbonation after 1, 3, and 5 years of natural carbonation was zero. During the process of natural exposure on the laboratory roof for 1 to 3 years, the strength increased slightly, but after 5 years of exposure, the strength had decreased slightly, which may be caused by the action of dry–moisture cycles during 5 years and fluctuations in temperature.

It was pointed out by Toru Kawai[75] that in an accelerated carbonation test with 100% CO_2 gas at a pressure of 0.4 MPa carried out for 40 h on 100 MPa super-high-strength concrete, the carbonation depth was zero compared to the carbonation of ordinary concrete, which reached 7.5 mm. The test results given by Gjorv[13] showed that after 1 year of artificial carbonation, the depth of carbonation of super-high-strength concrete with a water/cement ratio of 0.2 to 0.3 was also zero.

Even for ordinary concrete, it is possible to enhance its resistance to carbonation. Toshio Saito[5] stated that the addition of a durability-improving agent (ethylene glycol ether and amyl alcohol) to ordinary concrete with a water/cement ratio of 0.5 reduced the rate of carbonation. If the protective layer of concrete over the reinforcement were 4 cm, it would take about 800 years for carbonation of concrete to reach the surface of the reinforcement.

Table 9.5 Long-term strength and depth of natural carbonation of super-high-strength concrete

No.	Binder content (kg/m^{-3})	SF (%)	W/B	Fluidity (mm)		ρ_{28} (kg/m^{-3})	Compressive strength (MPa)							Depth of NC (mm)		
							Standard curing				28 d standard curing + natural curing on roof					
				Slump	Spread		3 d	28 d	56 d	90 d	1 y	3 y	5 y	1 y	3 y	5 y
1	600	10	0.20	110		2524	73.1	111.9	118.9	116.5	122.5	133.3	130.4	0	0	0
2	600	10	0.25	260	670	2518	66.6	96.8	105.2	104.1	120.6	121.8	121.1	0	0	0
3	600	10	0.30	280	800	2497	54.8	92.0	106.1	104.0	116.3	118.1	110.2	0	0	0

Note: SF = silica fume replacement; W/B = water/binder ratio; ρ_{28} = apparent density of concrete at 28 d; NC = natural curing. Naphthalene-based superplasticizer was added to all concretes at the same dosage. Long-term natural carbonation monitoring is still in progress.

9.5 ABRASION RESISTANCE OF SUPER-HIGH-STRENGTH CONCRETE

In many cases, abrasion-resistant concrete is required, such as for storage facilities for hard grain, troughs, workshop floors, and highway and expressway pavements,[105] especially in north European countries where studded tires used to prevent skidding in winter cause the road surface to wear quickly. Hydraulic structures, such as spillways, surge basins, and drains are also sensitive to hydraulic abrasion or erosion caused by sand grains in the water.

The abrasion resistance of concrete is dependent on many factors, such as the hardness of the aggregate, the water/binder ratio, the curing conditions, and age. In fact, it is directly related to the strength of concrete. Figure 9.7 illustrates the relationship of abrasion depth of concrete with its strength,[76] and it can be seen that for ordinary and high strength concretes, an increase in concrete strength gave improved abrasion resistance and the depth of abrasion decreased.

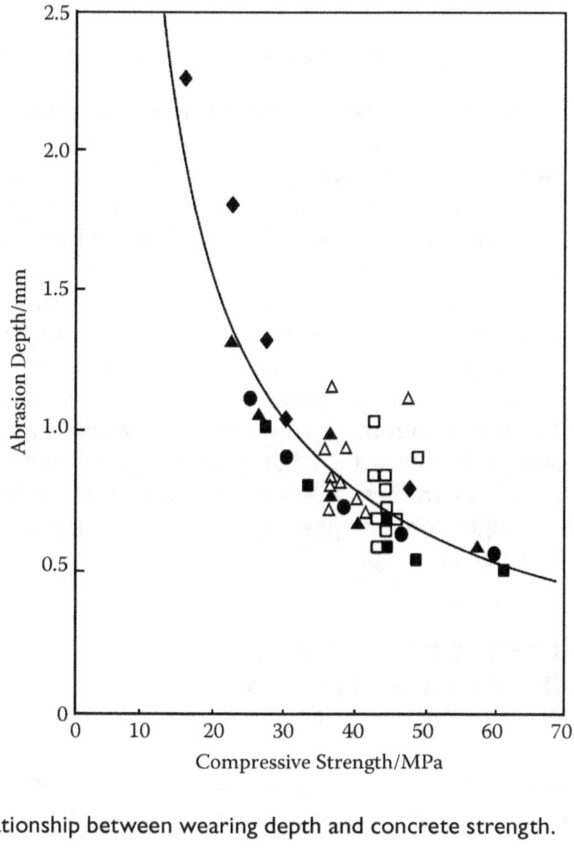

Figure 9.7 Relationship between wearing depth and concrete strength.

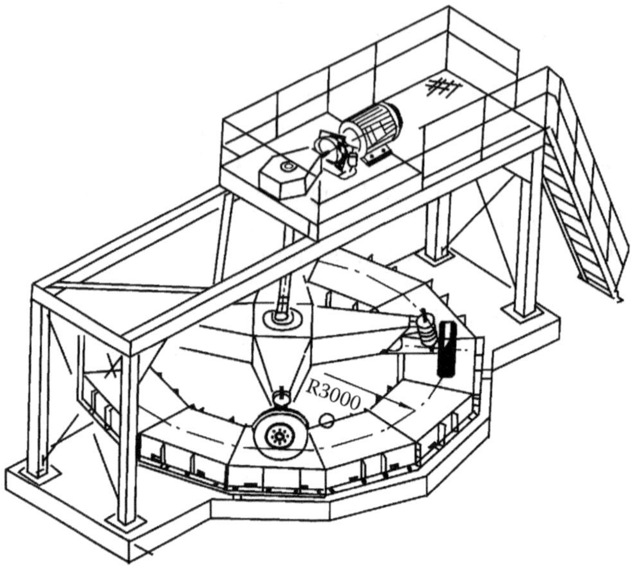

Figure 9.8 Norwegian testing machine for abrasion resistance of concrete.

In 1985, a full-scale test program on the abrasion resistance of an expressway pavement under heavy traffic exposed to studded tires with accelerated loading was set up in Norway (Figure 9.8). Test results are given in Figure 9.9.[13] In the high strength concretes category, an increase in concrete strength gave a linear decrease in the abrasion of concrete. If the concrete strength was raised from 50 MPa to 100 MPa, the abrasion of concrete was decreased by about 50%. When the concrete strength reached 150 MPa, the abrasion resistance of concrete was increased to the level of massive granite. In comparison to the typical asphalt pavement in Norway, the service life of a super-high-strength concrete pavement in an expressway was enhanced 10 times.

In northern European countries, where there is widespread use of studded tires, the service life even for high quality asphalt pavement is only 2 to 3 years. Based on the test results given above, some high strength and super-high-strength concrete pavements have been constructed in these countries (see Chapter 11).

9.6 COMPARISON OF THE DURABILITY OF ORDINARY, HIGH STRENGTH, AND SUPER-HIGH-STRENGTH CONCRETES

To compare the durability of ordinary, high strength, and super-high-strength concretes, Rollet et al.[77] made a systematic and comprehensive study of the

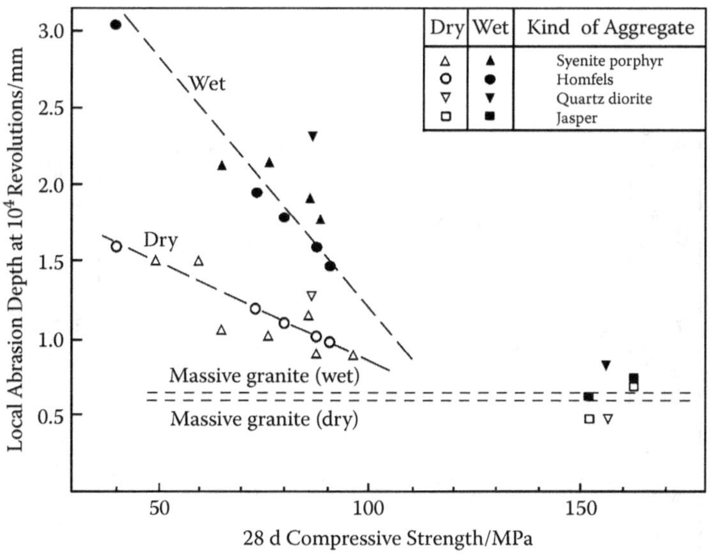

Figure 9.9 Relationship between abrasion resistance and compressive strength of high strength high performance concrete.[13]

durability of concretes ranging in strength from 20 MPa to 120 MPa and determined some of the physical and mechanical property indices of the various grades of concrete. Using similar mixing procedures and the same raw materials, 16 concrete mixes were prepared with slumps in the range 150 to 160 mm (Table 9.6). Eight mixes were made with an air-entraining admixture.

After drying the specimens below 60°C, the porosity was obtained by immersing the specimens in water under vacuum (called *water accessible porosity*) (Belgian Standard of Stone Material NBN·B10503). The results obtained are close to those obtained by mercury intrusion.

In this study, the following tests were carried out: (1) alkali silicate expansion (according to NF-B10503); (2) acid resistance (ϕ160 mm × 30 mm specimen soaked in 0.05 *M* lactic acid with a pH = 4.8 for 28 d and mass loss determined); (3) freeze/thaw test (according to NF-P16419 with a maximum of 300 cycles and determination of the relative dynamic modulus of elasticity); (4) sulfate resistance on a 20 mm × 20 mm × 160 mm prism of mortar sieved from the concrete and cured for 28 d, before soaking in an $MgSO_4$ solution (50g/L) and river water for 3 months, and expansion determined from the deformation of the specimen soaked in river water taken as 100%; (5) spalling (according to ASTMC672-84); (6) accelerated carbonation (φ110 mm × 110 mm specimen cured for 28 d, exposed in an accelerated carbonation box with 100% CO_2, relative moisture 60% for

Table 9.6 Mix proportions of concretes with different strengths (units: kg/m^3)

Grade	*C*	*FS*	*W/B W/(C+FS)*	*Coarse aggregate 4–14 mm*	*Fine aggregate 0.5–5 mm*	*Fine aggregate 0–0.5 mm*	*Limestone filler*	*SP (liquid)*	*A (liquid)*
B20	250		0.880	1033	439	276	100		
B40	400		0.536	1032	448	266			
B40FS	400	40	0.545	963	434	235			
B60	400		0.440	1063	468	266	50	4.5	
B80	450		0.351	1055	478	262	50	12.5	
B80FS	400	40	0.370	1092	476	272		11.6	
B100FS	450	45	0.323	1059	479	257		17.3	
B120FS	550	55	0.270	1003	479	237		30.2	
B20A	250		0.810	1012	430	277	100		0.5
B40A	400		0.493	997	433	259			2.2
B40FSA	400	40	0.545	907	408	215			0.4
B60A	400		0.475	970	435	235	50	4.5	0.45
B80A	450		0.358	991	448	292	50	12.5	0.5
B80FSA	400	40	0.345	1028	453	259		11.0	0.44
B100FSA	450	45	0.327	988	447	236		17.3	0.55
B120FSA	550	55	0.281	980	472	206		30.2	0.91

Source: M. Roller, Physical, Mechanical and Durability Characteristics of Various Compressive Strength Concretes (from 20 to 120 MPa), in Proceedings of CANMENT/ACI International Conference on Advances Concrete Technology, CANMENT, Athens, Greece, 1992.

Note: C = cement content; SF = silica fume dosage; W/B = water/binder ratio; SP = superplasticizer; A = air-entraining agent. For coarse aggregate, the alkali-active crushed limestone was used.

91 d and the depth of carbonation determined); (7) oxygen permeation (Cembureau[77] suggested using a 150 mm × 50 mm hollow cylinder cured at 20°C 100% relative humidity for 28 d, dried at 50°C for 6 d, and 20°C for a further day, absolute pressure of oxygen 0.4 and 0.5 MPa); and (8) a chloride ion penetration (AASHTO method). Test results are given in Table 9.7 and Figure 9.10. From these it can be seen that

1. It is well known that with a decrease in the water/cement ratio of concrete, the porosity and permeability of concrete decrease. As the compressive strength of concrete was increased, the penetration factor of oxygen, K_{O2}, decreased for concretes with strengths from 20 MPa to 70 MPa, then became constant up to 120-MPa concrete. Compared with non-air-entrained concrete, the O_2 permeability of air-entrained concrete was reduced, especially for low grade concretes.
2. In the case of accelerated carbonation, high strength concretes (over B60) and super-high-strength concretes gave excellent results. This is directly related to the water/binder ratio. The lower the water/binder ratio, the less the effect of accelerated carbonation.
3. The amount of chloride ion penetration is very large for ordinary concrete (B20, B40), where Q_{6h} (amount of Cl^- penetration in 6 hours) of B20 reached 24,850 C. The Q_{6h} value decreased rapidly with a reduction of the water/cement ratio and an increase of concrete strength. This value was lowered to 19 C for the B120FS concrete. For concretes of the same grade, such as B40 and B40FS, B80, and B80FS, addition of silica fume greatly decreased the chloride ion penetration. The higher the compressive strength, the smaller the penetration coefficient of chloride ion and the greater is T_0, the delay in the time at which the Cl^- ion penetrates. In short, the higher the strength, the smaller the porosity and the lower the amount of liquid in the pores, and the more difficult it will be for the chloride ion to penetrate.
4. Addition of silica fume (compare B40 with B40FS and B80 with B80FS) can effectively prevent the alkali–aggregate reaction. Usually, silica fume or other fine mineral admixtures must be added to obtain super-high-strength concrete. Such concrete has excellent impermeability and it is difficult for water to penetrate; thus without the presence of water, alkali–aggregate reaction will not occur.
5. Concrete is not an acid-resistant material, but in a weakly acidic environment, it still has the ability to resist acid action. Under the conditions studied, an increase in strength gives better acid resistance. Whereas the mass loss for the ordinary concrete was 8.7%, for the high strength concrete it was 6%, and for super-high-strength concrete, 5%.
6. The results of the $MgSO_4$ corrosion test correlated well with the strength of concrete, an increase of compressive strength increasing the corrosion-resistant ability and reducing the expansion (Figure 9.10[f]).

Table 9.7 Durability indexes of concretes of different grades

Grade	$R_{28\ d}$ (MPa)	Porosity (%)	Spacing factor (μm)	O_2 penetration ($10^{-18}m^2$)	Dcarb at 91 d (mm)	E_{AS} (at 6 months) (%)	ML of 28 d acid corrosion (%)	$MgSO_4$ corrosion (%)	Chloride ion diffusion			Frost resistance			Mass loss from spalling ($g \cdot m^{-2}$)		
									Q_{6h}/C	D ($10^{-6} \cdot cm^2 \cdot S^{-1}$)	t_0 (h)	Number of cycles	DF/%	E ($\mu m \cdot m^{-1}$)	20 cycles	40 cycles	50 cycles
B20	28.0	14.3	840	31.2	55.0	0.0073	8.7	1043	24850	671.0	4.4	34	0	7230	15002		
B40	50.8	10.1	760	14.2	4.0	0.0130	6.3	183	12470	253.0	15.1	56	43	3730	7558		
B40FS	56.9	9.7	797	10.5	10.1	0.0057	5.8	120	452	123.0	32.6	148	54	4670	6477		
B60	67.9	7.3	1134	4.7	0.6	0.0198	6.2	107	689	127.0	33.7	300	100	−7	4464		
B80	83.9	4.3	899	3.1	0.1	0.1211	5.2	103	178	45.0	13.0	300	100	−153	1078	2174	2817
B80FS	94.6	3.7	758	2.4	0.1	−0.0004	4.8	30	52	17.1	514.0	300	100	−159	1408	2494	2975
B100FS	101.7	3.4	678	1.9	0.1	−0.0023	4.8	26	29	5.4	736.0	300	100	−235	134	266	372
B120FS	113.3	3.1	817	2.6	0.1	−0.0042	4.9	20	19	4.8	1447.0	300	100	−212	50	64	116
B20A	27.0	17.5	375	63.2								300	100	−23	1684	2214	2570
B40A	38.5	14.0	313	32.4								300	100	−19	444	528	580
B40FSA	45.5	13.8	301	22.8								300	100	−26	624	708	754
B60A	51.2	11.6	275	13.9								300	100	−61	146	222	272
B80A	71.3	5.3	371	2.3								300	100	−110	76	108	
B80FSA	83.7	5.1	327	1.7								300	100	−87	630	1492	
B100FSA	91.7	3.1	339	2.0								300	100	−110	518	1204	
B120FSA	103.0	3.1	375	2.5								300	100	−118	318	1038	

Note: $R_{28\ d}$ = 28 d compressive strength (φ160 mm × 320 mm specimen); D_{carb} = depth of carbonation; E_{AS} = expansion rate of alkali–silica reaction; ML = mass loss; DF = durability factor for frost resistance; E = expansion rate.

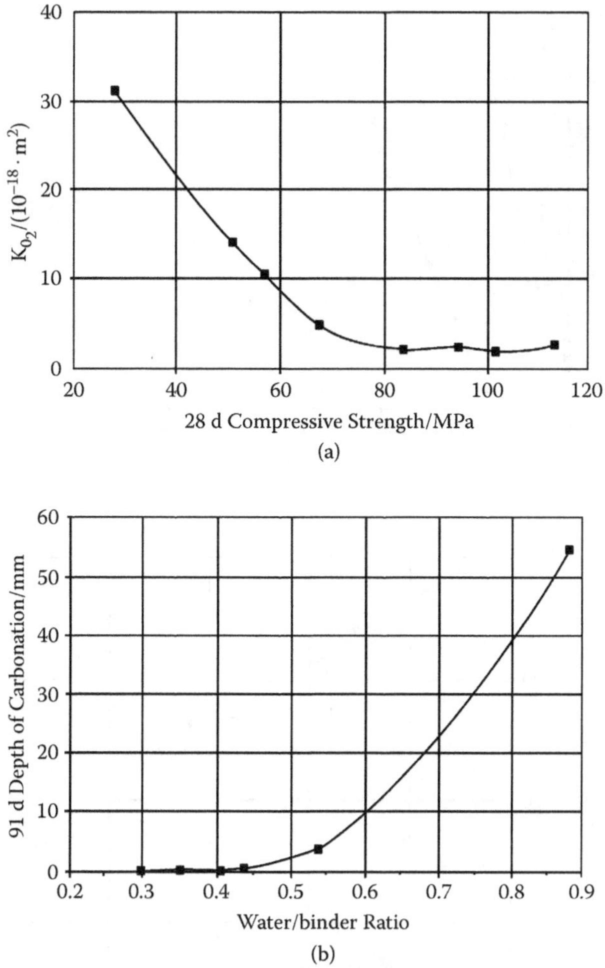

Figure 9.10 Relationship between durability indexes of concrete and compressive strength of concrete or time: (a) O_2 penetration; (b) carbonation resistance; (c) relationship between chlorine ion diffusion and time; (d) diffusion coefficient of chlorine ion; (e) starting time of chlorine ion diffusion; (f) acid corrosion resistance; and (g) sulfate corrosion resistance.

7. The frost resistance test showed that concrete with a higher strength has higher frost resistance. Concretes from B60 to B120 had excellent frost resistance. Regrettably, this test was not carried out to the end so the limit of the frost resistance of the high strength and the super-high-strength concretes remains unknown. Meanwhile, ordinary concrete (B20, B40, and B40FS) was rapidly damaged by frost, so for these concretes, air-entraining agents should be added to provide good frost resistance.

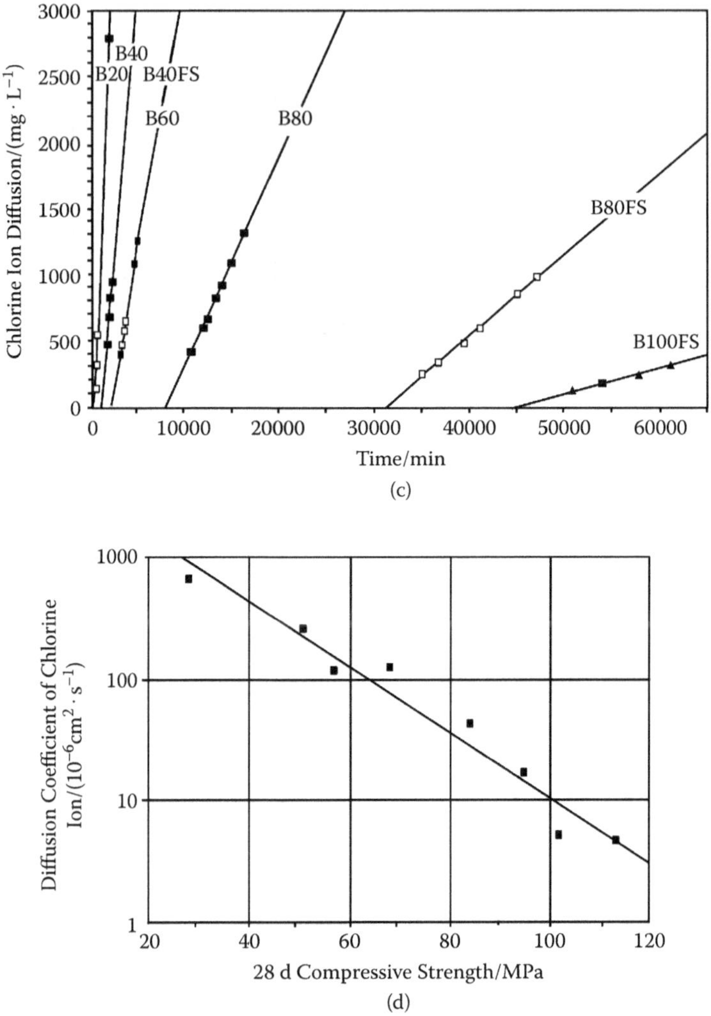

Figure 9.10 (continued)

8. According to the Swedish Standard on spalling (SS137224) and Canadian Regulations, after 50 freezing and thawing cycles, only concrete with a mass loss less than 1000 g/m² can be defined as high quality, spalling resistant. It can be seen from Table 9.7 that without an air-entraining agent, only B100FS and B120FS, that is, the super-high-strength concretes, obtained excellent spalling resistance. In concretes with an air-entraining agent, except for B20A, the spalling resistance of concretes B40A, B40FSA, B60A, and B80A was improved. What is unexpected is that the high strength and super-high-strength concretes

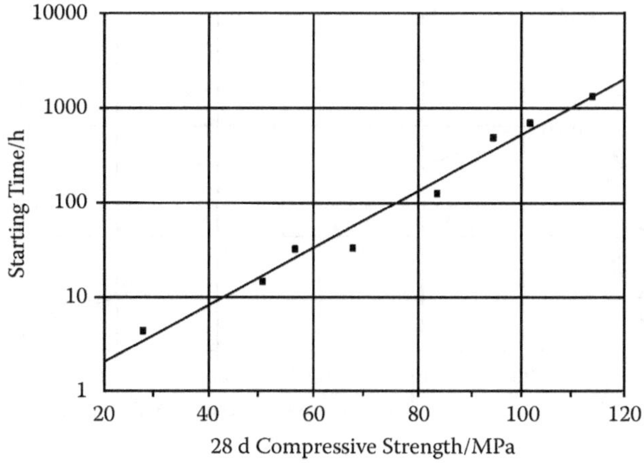

(e)

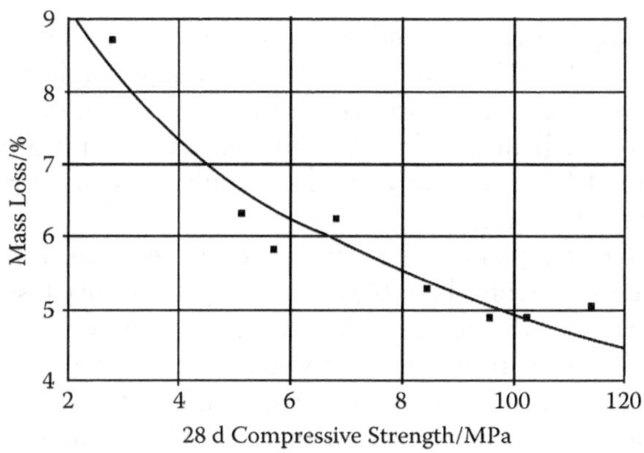

(f)

Figure 9.10 (continued)

B60FSA, B100FSA, B120FSA, and so forth gave undesirable results, which were explained by the researchers as due to the air bubbles in the very compacted concrete which may be the weak points of the concrete and the starting points for the sudden fracture. The author suggests that for super-high-strength concrete, it is not necessary to add an air-entraining agent; from the viewpoint of frost resistance and spalling resistance, such defects can be avoided.

After this study on the physical and mechanical properties and the durability of ordinary, high strength, and super-high-strength concretes, the researchers further identified the advantages of high strength and

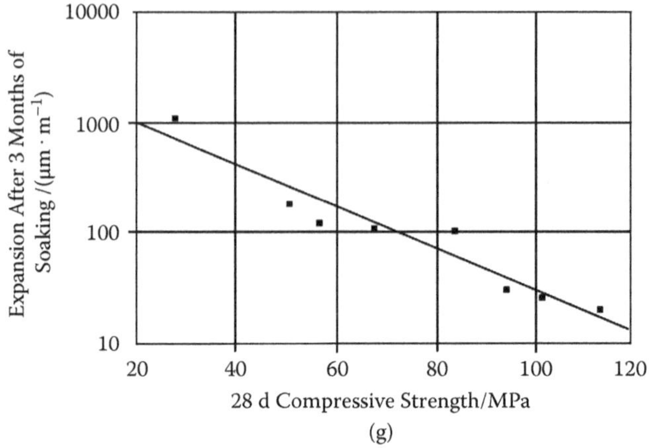

Figure 9.10 (continued)

super-high-strength concretes. Judged by any durability index, super-high-strength concretes B100FS and B120FS containing enough silica fume and super-plasticizer, and with a water/cement ratio less than 0.33, can be considered to have excellent durability, while concretes with a water/cement ratio less than 0.4 have relatively good durability. The author agrees with the researchers' viewpoint.

Finally, these researchers gave a comprehensive evaluation (Table 9.8) of results. It can be seen from the table that by any durability test, super-high-strength concretes are the best and the ordinary concrete, B20, is the worst.

However, the author thinks that concrete made in the laboratory is different from that used in building construction, and the difference may be large. Therefore, during construction there should be intensive control of procedures. An improvement in construction technology to bring it close to or identical to the quality used in laboratories is the key problem to be solved by researchers in materials science and building construction. The author thinks that solving this problem is an achievable goal.

Table 9.8 Classification of the concretes by durability index

Class of concrete	*W/B*	*28 d Rcom/ MPa*	*Resistance to:* Carbonation	*Cl-ion*	*AAR*	*Acid attack*	*Seawater corrosion*	*Sulfate corrosion*	*Frost and thaw cycle*	*Spalling*	*Complex class*
B20	0.880	28.0	8	8	5	8	4	8	8	8	8
B40	0.530	50.9	6	7	6	6	6	7	7	5	Uncertain, it depends on what types of durability are studied
B40FS	0.545	56.9	7	5	4	5	4	6	6	5	
B60	0.440	67.9	5	5	7	6	7	4	1	5	
B80	0.351	83.9	1	4	8	4	6	4	1	3	
B80FS	0.370	94.6	1	3	1	1	1	1	1	3	3
B100FS	0.323	101.7	1	2	1	1	1	1	1	1	1
B120FS	0.270	113.3	1	1	1	1	1	1	1	1	1

Note: AAR = alkali–aggregate reaction.

Chapter 10

Test methods

Modern concrete has been used for about 170 years, though large volumes of high strength concrete only came about in the last 20 to 30 years, while the application of super-high-strength concrete has only just started. Many of the testing methods and standards that have been set up are based primarily on ordinary concrete, with little consideration of the characteristics of high strength and super-high-strength concretes. Application of existing test methods for ordinary concrete adapted for testing high strength and super-high-strength concretes is obviously inappropriate. Therefore, in this book, some of the experimental results from literature data and the author's research are presented for discussion.

10.1 TESTING METHOD FOR FLUIDITY OF THE CONCRETE MIX

From a rheological viewpoint, a concrete mix behaves similarly to a Bingham body so its rheological properties can be determined by a rotary viscometer.[19] Gjorv[13] studied the rheological behavior of a high strength concrete mix using a rotary viscometer (see Chongqing Institute of Architecture and Engineering and Nanjing Institute of Technology[19] and Feng Naiqian[78] in Chinese with its full citation). However, the most suitable material to use in a rotary viscometer is a paste containing fine powders, as the viscometer is not well adapted to a concrete mix containing coarse and fine aggregates. At present, there is no apparatus that can determine the rheological performance of a concrete mix, a complicated system composed of a mixture of solid particles and an aqueous phase, viz., coarse and fine aggregates, cement and fine additives with large differences in particle sizes suspended in water.

The widespread method for measuring the fluidity of a concrete mix is the traditional slump measurement. However, it is difficult to express the fluidity of different concrete mixes this way, as at times concretes with the same slump have quite different properties.

Table 10.1 Comparison of the fluidity of the concrete mixes

No.	Binder content (kg/m³)	Binder composition (%) Cement	SF	Water content (kg/m³)	W/B	Sand content (kg/m³)	Crushed limestone content (kg/m³)	Fluidity (mm) Slump	Spread	Compressive strength (MPa) 3d	28d
1	600	90	10	150	0.25	720	1080	220	570	65	103
2	600	90	10	150	0.25	720	1080	215	350	62	110

Note: SF = silica fume; W/B = water/binder ratio; the water-reducing agent for mix No. 1 is S-20 and for mix No. 2 is FDN.

Some researchers have suggested utilizing an L-shaped slump tester and determining the L-shaped slump and the distance of flow of a concrete mix to estimate fluidity. However, such an L-shaped apparatus is more complicated than the usual slump cone and the mix is influenced by the resistance of the four cones and two grooved walls. The author thinks that if the spread of the mix cone was measured at the same time as the slump in the traditional measurement of slump, the fluidity of the concrete mix could be estimated. For example, in Table 10.1, the slump of mix No. 1 is close to that of mix No. 2 but the spread of mix No. 1 is much greater than that of mix No. 2. Consequently, the fluidity of mix No. 1 is better than that of mix No. 2. The principle of such a method is similar to that of the L-shape tester, where the mix slumps down and then flows downward due to gravity, but the difference is that the flow of the mix in an L-shape tester is only in one direction, whereas in the slump cone test, the flow is radial. The important thing is that measurement of both the slump and spread of the mix is convenient and can be implemented simply and the data can be compared with that obtained by the traditional method given in the standards of different countries. Therefore, the author recommends this method and further, if the time from the start of the slump to the point when the spreading cone has spread a certain distance is measured, the speed of the flow at a particular time can also be determined and this method further improved. In addition, there is a popular new application of the slump cone, where the cone is turned upside down and filled with a concrete mix and the time to empty the cone measured giving an effective way to measure the fluidity of the concrete.

10.2 THE STANDARD AGE FOR STRENGTH MEASUREMENT

According to the Chinese Standards GBJ 81-85, the strength of concrete at 28 d is the normal age for determining strength. This age was specified many years ago and is based on ordinary concrete, which was the situation at the time, when the extent of construction was small, time for construction was short, and the concrete expected to bear the design load by completion of the project. The control strength of concrete was required in a short time and 28 d age was suitable. Besides, for ordinary concrete, strength develops rapidly within 28 d, with slower strength development after that.

For high strength and super-high-strength concretes, which are mostly used in very large or giant projects, the construction duration is usually long, 1 or more years, so the design load is imposed on the structure after 1 or more years. For high strength and super-high-strength concretes, the concrete can bear the construction load at 3 d to 7 d strength, allowing later age strength to be used as the design strength to fully utilize the potential strength.

Usually, to obtain high strength and super-high-strength concretes, superfine mineral admixtures called *auxiliary binders* are added. Some additives, in particular fly ash, slag, and so forth, have a low rate of reaction in concrete, and only in later ages does the contribution of the pozzolanic effect to strength develop. Even silica fume, an additive with an early secondary reaction, still shows strength growth in later ages. Ignoring this strength development is undoubtedly a waste of the resource.

Due to the two reasons given above, the author suggests that the 90 d strength should be taken as the standard strength for high strength and super-high-strength concretes to adapt to the new situation. In fact, for hydro dam concretes, which have mostly low strengths, the 90 d strength has been taken as the standard strength for a long time, as hydraulic construction usually occurs over a very long duration. According to Wang Shaodong,[27] in the American standard of concrete construction design, 56 d strength has been taken as the standard strength value. We should follow such a trend in development.

10.3 SHAPE AND DIMENSION OF THE TEST SPECIMEN

Nowadays, in Europe and other countries as well as in China, cubes are used for determining concrete compressive strength, but in North America, cylinders are used. In China, the standard dimensions of the concrete specimen are 150 mm × 150 mm × 150 mm, whereas for countries where a cylinder is used, the standard dimensions are φ150 mm × 300 mm. In these standards, a 100 mm × 100 mm × 100 mm cube or a φ100 mm × 200 mm cylinder are permitted for the test. The strength obtained is multiplied by a conversion coefficient of less than 1 to convert to the strength of a standard specimen.

Usually the strength value determined with a cylindrical specimen is less than that with a cube. This is understandable because for the cylinder specimen, the height/diameter ratio is 2, while for the cube specimen the height/width ratio is 1. The ratio for the cylinder specimen is much higher and the restriction effect of both ends on the center is reduced, so consequently, the strength of the specimen will be reduced.

Which kind of specimen should be used for super-high-strength concrete? The author thinks that the 100 mm × 100 mm × 100 mm cube is suitable. In a cylinder, one of the surfaces for compression is cast and unlikely to be smooth and another material is needed for leveling or capping or the surface must be ground. For super-high-strength concrete, it is difficult to find a capping material to match it and capping is troublesome; the capped surface will not necessarily be exact or satisfactory. Super-high-strength concrete has high abrasion resistance, so it is difficult to grind smooth. For the cube specimen, such as that used in China, there is no problem with loading the cast surface.

The reason for suggesting the use of 100 mm × 100 mm × 100 mm cubes instead of the 150 mm × 150 mm × 150 mm cubes specified in the present standard is as follows:

1. The 150 mm × 150 mm × 150 mm cube specimen is cumbersome. The preparation, lifting, and testing of the specimen is hard physical work, therefore laboratory professionals and some researchers prefer the smaller specimens to the large cumbersome ones. In the 1950s and 1960s, within the China Design Standard, the 200 mm × 200 mm × 200 mm cube specimen was specified as the standard; however, in laboratories and on building sites, it was rarely used and the 100 mm × 100 mm × 100 mm cube was used instead. The difficulty of handling the large 200 mm × 200 mm × 200 mm cubes was one of the reasons why the 150 mm × 150 mm × 150 mm cube was taken as the standard specimen. Meanwhile, use of the smaller specimen will reduce material consumption and it provides flexibility in mix proportion variations.
2. Elimination of a conversion factor. In the China Design and Construction Guide for the high strength concretes (HSCC-99), it is specified that when a 100 mm × 100 mm × 100 mm cube is used, the compressive strength $f_{cu.10}$ obtained should be multiplied by the conversion factor given in Table 10.2 to convert it into equivalent data for the standard 150 mm × 150 mm × 150 mm cube.

   It can be seen from Table 10.2 that the strength value is reduced linearly by a factor of 0.1/100 MPa with an increase in strength. If such a rate is used for super-high-strength concrete, then the reducing factor for 150-MPa concrete should be 0.85.

   There is some difference of opinion on this value; for example, Qian Xiaoqian and Zan Shulin[79] have shown from test results with large and small, rather precise specimens with a smooth surface, that

Table 10.2 Conversion factors (deducting factor) of compressive strength of concrete cube K

$f_{cu.10}$ (MPa)	K	$f_{cu.10}$ (MPa)	K
≤55	0.95	76–85	0.92
56–65	0.94	86–95	0.91
66–75	0.93	≥96	0.90

Source: High Strength Committee and China Civil Engineering Association, *Guide to Structural Design and Construction for High Strength Concrete*, 2nd ed., Beijing: China Building Industry Press, 2001 (in Chinese).

a conversion factor for C60–C80 concretes of 0.95 was obtained, which means the conversion factor does not decrease with an increase of strength. Wang Yurong,[81] however, gave a conversion factor for strength of concretes with compressive strengths between 95 and 113 MPa as 0.9853.

The study by Sviridov[41] showed the statistic relationship between the specimen dimensions and the homogeneity of the concrete. For super-high-strength concrete, its mechanical properties are highly consistent, compared to standard specimens, where dimensions differing from the standard all have a dimension factor of about 1. Such a consistency in strength is due to the high homogeneity of the super-high-strength concrete.

Herd has studied the relationship in strength between cylinder and cube specimens of concrete with strengths between 50 and 100 MPa and the results are given in Table 10.3. It can be seen from the table that the conversion factor between a cube with a side length of 100 mm and one with 150 mm is 0.99. In the same literature, the conversion factor between a cube with a side length of 100 mm and a φ 150 mm × 300 mm cylinder specimen is given. Different data show that the conversion factor increases with the concrete strength (Table 10.4), contrary to that given in Table 10.2.

Table 10.3 Conversion factor for strength between shape and dimensions of high strength and super-high-strength concrete specimens

Shape and dimension	*100-mm cube*	*150-mm cube*	*200-mm cube*	*Φ 150 mm × 300 mm cylinder*
100-mm cube	1	0.99	0.95	0.82
150-mm cube		1	0.95	0.83
200-mm cube			1	0.87

Table 10.4 Conversion factors by Smeplass[13]

Concrete strength (MPa)	*Cube of 100 mm × 100 mm × 100 mm*	*Cube of 150 mm × 150 mm × 150 mm*
66.3	1	0.75
97.0	1	0.77
115.4	1	0.83

Source: Odd E, Gjorv, High Strength Concrete, in *Advances in Concrete Technology*, CANMENT, 1992.

The author has used a 500 t press to measure cube specimens with strengths of 120 MPa and a side length of 100 mm and 150 mm (across the ground end surfaces) and obtained a conversion factor of 0.98.[82]

Due to differences in raw materials, preparation of specimens, testing machines, and the quality of the operators, the results obtained and the conclusions drawn are not consistent and the dispute will continue. The author suggests that the dimensions of the standard specimens used for design should be unified with those used in laboratories and on building sites. An inappropriate conversion factor due to specimen dimensions will lead to waste of resources or even potential danger.

3. In the design of mix formulations for buildings, the maximum particle size of coarse aggregate used for concrete has decreased, due both to increased strength in the concrete used and greater congestion around the reinforcing bars, which has forced particle size reduction of the coarse aggregate to ensure sufficient workability of the concrete mix. In addition, in the blasting which occurs during mining and crushing, the probability of the presence of microcracks developing is more likely in the larger aggregate size. Therefore, reduction of maximum particle size of coarse aggregate for high strength and super-high-strength concretes to 10 to 14 mm has been suggested by many experts to be applied worldwide. Thus, with this reduced maximum particle size of coarse aggregate, for example, concrete cube specimens of 100 mm size, prepared from coarse aggregate with a maximum particle size of no more than 20 to 25 mm and appropriate grading, could be used and the macrostructure of a homogeneous stacked agglomerate could be observed by cutting it open in halves. The 100 mm × 100 mm × 100 mm cube specimen has been applied as a standard that is representative. Of course, dam concrete is an exception, where the maximum particle size of aggregate reaches 80 to 150 mm.
4. In the case of super-high-strength concretes, if the 150 mm cube specimen is used, the capability of the crushing machine is inevitably increased. For example, in super-high-strength concrete with a strength of 150 MPa, the breaking load should be 338 t. Considering a 20% margin, the minimum capacity of the testing machine required is 406 t.

If no such machine is available, then a 500 t press should be installed in a laboratory. If a 100-mm cube specimen is used with a minimum breaking load of 150 t, then only a 200 t press is acceptable. While it is difficult for all construction companies and ready-mix concrete plants to install 500 t presses, the 200 t press is present in many large plants and construction enterprises. Therefore, adopting a cube specimen with a side length of 100 mm as the standard will promote widespread acceptance of super-high-strength concrete and advance the development and acceptance of new technology both in China and the world. There are already some countries in the world where a 100-mm cube is the standard, for example, in the Norwegian Design Standard for Concrete Structures (NS 3473E),[83] the strength of a 100-mm cube specimen is taken as standard for concrete with a max strength grade of C105.

10.4 TESTING TECHNIQUES FOR SUPER-HIGH-STRENGTH CONCRETES

10.4.1 Mold

As described above, super-high-strength concrete has low plasticity and little plastic deformation, so the shape, dimension, and surface smoothness of the specimen, as well as the parallelism of the two load-bearing surfaces, play an important role in the test results. During the testing process of ordinary concrete, the stress redistribution can be realized through plastic deformation to distribute the stress evenly. Therefore, in this respect, ordinary concrete is superior to high strength and super-high-strength concretes. Consequently, super-high-strength concrete without its plastic deformation has a high requirement for the crushing surfaces of the specimen, as well as the load-bearing plate of the testing machine, to be parallel. So, before testing super-high-strength concrete, the load-bearing surface of the specimen is ground. Such work is simple. Fine emery powder is sprayed onto a thick steel plate with a smooth surface and enough stiffness. The specimen is placed on the surface and moved in circles with hand pressure and checked periodically to ensure there is close contact between the load-bearing surfaces of the specimen and the load-bearing plate of the testing machine which is ground flat. If there is a point not in close contact, the grinding is repeated until there is close contact between both surfaces. The thickness of the load-bearing plate of the testing machine should be sufficiently stiff in order to avoid any influence caused by the smoothness of the pressing head of the testing machine. The steel should be hard enough. The internal surface of the mold should be clean and smooth, giving a specimen with a smooth plain surface to reduce final grinding.

The specimen mold should have high stiffness and not be subject to deformation. Its assembly should guarantee that it is cubic, which can be observed visually and the molds should be checked one by one. Leaning specimens will give rise to eccentric axes, even if the upper and lower surfaces are parallel, as on the testing machine when the lower surface is centralized; the upper will not be centralized giving a lower result.

Non-parallel, load-bearing surfaces will also lead to unsatisfactory results, although a ball support can be used to ensure close contact of the upper load-bearing surface with that of the concrete during testing, as any offset load may impact on the specimen.

10.4.2 Preparation of the specimen

The cement used for testing should be stored hermetically in a plastic barrel to avoid the effects of moisture, thus ensuring the quality of the cement remains unchanged during long-term storage. For example, the author has stored a Kunming 625 cement in such a way for 11 months and retesting showed (Table 10.5) that there was no reduction in activity of the cement, with even a little increase, confirming that cement can be stored this way for at least 11 months.

The bulk materials for testing, such as sand and crushed stone, should be stored in an indoor silo to avoid changes in moisture content and contamination by dust or foreign materials. To implement a key project of the Natural Science Foundation of China more than ten such silos were installed in the Laboratory of Concrete Technology at Chongqing University, to avoid storage in the open air.

The concrete should be mixed in a high shear concrete mixer. Due to the low water/binder ratio of a super-high-strength concrete mix along with the large amounts of powder in the mix, the mix can be rather thick. During casting, the material should be added to the mold held on a vibrating table or added in layers with suitable compaction. It is not advisable to fill the mold with a single addition of mix, otherwise air trapped in the molded mix and the mold cannot be removed and pores will be formed

Table 10.5 Activity of cement after long-term storage[a]

First grade determination				*Grade determination after 11 months*			
3 d		*28 d*		*3 d*		*28 d*	
Bending	*Compressive*	*Bending*	*Compressive*	*Bending*	*Compressive*	*Bending*	*Compressive*
7.09	30.7	9.49	74.9	7.93	43.0	10.69	76.8

Note: Units = MPa.

[a] (GB 177-85) China.

in the specimen. After casting and finishing of the surface, the specimen should be moist cured after initial set and covered with a plastic film.

Two or three more sets of specimens should be prepared as a reserve.

10.4.3 Compression test

Half a day before compression testing, the specimens should be taken from the normal curing room, dried indoors, and the load-bearing surfaces smoothed. The compression testing is carried out after a quality check and dimensional measurement. The testing machine should be calibrated. The lower loading plate (there should be a mark at the center of the plate) should be centered and the dust cleaned off the load-bearing surfaces of the specimen and both load-bearing plates of the testing machine. The contact between specimen and load-bearing plates should be checked at both corners of the diagonal by pressing with both hands alternatively to check if there is any movement as well as ascertaining whether the specimen can rotate on the plate. If there is any shaking or rotation during the check, which shows that there is not close contact between the specimen and the loading plate, the surface of the specimen should be smoothed again. If this cannot be done, the reserve specimen may be used.

It is important to center the specimen, as any eccentricity will lower the strength value so a true strength value may not be obtained. To make centering easier, at the center of the load-bearing plate surface, there is a square mark of 100 mm × 100 mm with a shallow slot carved into it. Occasionally, the bearing surface of the specimen is not an accurate square; it should be adjusted in two perpendicular directions. After locating the specimen, the upper loading plate is installed, with close contact between the plate and specimen.

During compression testing, the oil supply for the machine should be uniform and the pointer of the meter should move uniformly; as the rate may be faster than that for ordinary concrete specimens due to the long process of loading (2~3 times more than ordinary), the test lasts longer.

The normal breakdown of the specimen is as follows:

1. There is no sound emitted during compression testing, and sudden blasting will occur upon breakdown. If a sound does occur during testing, particularly a loud sound, it indicates local damage has occurred. Such a failure is abnormal and the concrete has low strength. Sometimes, there is a low sound from shattering at the surface but this is unlikely to cause any serious problem.
2. When the specimen breaks, it forms two opposing tetrahedral cones, but due to the extremely high strength of the super-high-strength concrete, the two cones left are usually rather small with other parts lost by shattering. Failure on one side, or by splitting, as well as failure along the diagonal are all abnormal.

In the case where the difference of strength values of specimens from a group exceeds that specified by the standard, the data obtained should be considered ineffective. If the scattered data are caused by reason of preparation or compression test procedure, it is recommended that the test be repeated with spare specimens for retesting to obtain a true result.

10.5 TEST METHODS FOR OTHER PROPERTIES

Due to the high level of compaction, high brittleness, and high strength characteristics of super-high-strength concrete, test methods adapted from ordinary concrete are not suitable for super-high-strength concrete. If these methods are used for monitoring the properties of super-high-strength concrete, the following may occur:

1. Wrong results can be obtained.
2. Test results cannot be judged or quantified.
3 Extreme consumption of time and manpower.

For example, we tested the shrinkage of super-high-strength concrete using the standard method for ordinary concrete (GBJ 82-85) and came to the conclusion that the shrinkage at 180 d was no more than 500×10^{-6},[84] which is similar to that of ordinary concrete. The study on self-shrinkage concluded that it made up a large part of the total shrinkage. The main concern is that the shrinkage occurs mainly in the initial 3 d, but according to GBJ 82-85, the initial length is to be measured 3 d after mixing, so the shrinkage in the first 3 d is neglected. Wang Yongwei[54] set the initial length when the mold was removed and the strength reached 3 to 5 MPa. The maximum total shrinkage of super-high-strength concrete reached 1579×10^{-6}, and the 3 d shrinkage made up 65.5% of the total. The self-shrinkage at each age made up about half of the total (see Table 8.8).

It is obvious that much of the shrinkage is omitted if GJB-82-85 is used. Many concrete specialists suggest that the shrinkage of super-high-strength concrete should be determined from the point of initial set of specimen in the mold. Though the method suggested by Wang Yongwei[54] is not perfect, the shrinkage determined by his method may have some loss compared with that determined after the end of initial set (the specimens were not demolded, which can resist the shrinkage of concrete), but his method shows many improvements and is simple in operation. There is no national standard or regulation at present for determining the shrinkage of super-high-strength concrete but research is continuing.

In another example, in testing the permeability of super-high-strength concrete using the present GBJ 82-8 method on the HS-40 permeability-testing machine, all results gave grades of permeability resistance over P40.

After the specimen was cut open, little or no leakage was found (Table 9.1), but the degree of resistance could not be judged or compared. Even when the chloride ion permeability test was used, it was still difficult to distinguish which formulation was better. As shown in Figure 1.1, when the water/binder ratio is less than 0.3, the electricity passing through a concrete containing silica fume is zero.

Yet in a further example, if the slow freeze/thaw test in the Chinese standard GJB82-1985 is used for examining super-high-strength concrete, several years would be needed to attain the 25% strength loss and mass loss of more than 5%. Even if the rapid freezing/thaw test is applied, a long time is needed. If one test cycle takes 3 h, the maximum number of cycles that can be conducted is 8 per day. Testing for 1200 cycles would then require no less than 120 d. As shown in Table 9.3, C100 super-high-strength concrete has a reduction in relative modulus of elasticity by 1.8% to 3.1% with a mass loss of zero after 1200 cycles of freezing and thawing. For a mass loss of 5% and a reduction in the relative modulus of elasticity by 40%, how long is needed? Of course, in this standard, it is specified that concrete passing 300 cycles of testing is considered excellent in frost resistance, and the testing may be stopped. In many cases, 300 cycles are not enough, and higher frost resistance is required.

Summarizing the results above, the author suggests that as the strength grade of concrete is increased, the requirements for testing methods should be improved accordingly, scientifically, and in specification. The testing machines should be very accurate, and the knowledge and responsibility of the laboratory professionals should be improved. Only in this way can new tests be developed for super-high-strength concrete so accurate results can be obtained. The study of test methods and the compilation of standards for super-high-strength concrete should be promoted, and research efforts into super-high-strength concrete should increase in order for its acceptance and use to become popular.

Chapter 11

Applications, trials, and prospects

11.1 APPLICATIONS AND TRIALS FOR SUPER-HIGH-STRENGTH HIGH PERFORMANCE CONCRETE

Due to the hard work needed to prepare super-high-strength concrete, and especially the strict technical requirements needed for quality construction, there has been no extensive study on the strength and deformation behavior of super-high-strength concrete subjected to different static and dynamic loads. Structural engineers do not yet fully understand the specific requirements for designing with super-high-strength concretes used in engineering. What is more important is that in most countries, the standards have not kept pace with new developments and there have been no related standards for structural design and construction developed. Only in Norway and Germany have suitable standards been developed. Norway has a standard, NS3473E *Design Standard of Concrete Structures* published in 1992, which provides the design norm for super-high-strength concrete with a maximum of C105. In 1995, the German Association of Reinforced Concrete published the *Instruction to Prepare High Strength Concretes* with the highest grade of concrete strength C115. Due to the lack of standards, the application and acceptance of super-high-strength concrete are limited.

In spite of this, enterprising companies, researchers, engineers, and technicians have been applying super-high-strength concrete in practical projects and excellent success has been achieved. The following are examples of application and trials of super-high-strength concrete in engineering throughout the world.

1. In 1959, Beijing Construction Bureau 2nd Co., Beijing Research Institute of Construction Engineering, Beijing Design Institute of Industrial Construction, and Beijing Research Institute of Architecture jointly carried out an experimental study developing Grade 1000 (about 100 MPa) super-high-strength concrete, and they manufactured a prestressed trapezoidal roof truss with a span of 18 m. Following this, 6000 m^2 of industrial buildings have been completed.[39]

The cement used in this project, was 800 grade (about 80 MPa) and manufactured by a secondary grinding clinker from Jiangnan Cement Plant with a vibrating mill to a specific surface of 600 m^2/kg. The cement content of the concrete was 750 kg/m^3 with a water/cement ratio of 0.26. The concrete was specially designed with a maximum particle size of crushed stone of 10 mm, the sand percentage 26%, with the addition of 0.3% of a lignosulfonate plasticizer. The stiffness value of the concrete was 30 to 60 s and the concrete mix was compacted after casting with a vibrating poker with additional compaction given by vibrating the external side of the mold. The roof truss was manufactured as follows: the chord and web members were manufactured first at the building element plant, and then the truss was assembled from these members at the building site and installed. Thirty years after its completion, investigations of the truss found no trace of damage on the super-high-strength concrete truss, where the surface was smooth and clean with the edges and corners intact.

2. The BFG high-rise office building in Frankfurt, Germany, built in 1993, is 186 m high with 47 stories and 4 stories of basement. The design-strength grade of concrete for some of the columns and walls in the lower stories was C85 with a maximum cross section of the columns 1.2 m × 1.2 m. The mix proportions of the concrete used were as follows: cement content of 450 kg/m^3, silica fume of 35 kg/m^3, water/binder ratio of 0.32, natural sand, 0 to 2 mm, of 660 kg/m^3, gravel, 2 to 16 mm, of 1170 kg/m^3, superplasticizer of 12 L/m^3, and a set retarding agent of 1.8 L/m^3. The spread of the concrete mix was 500 to 600 mm, which was suitable for placement by pumping. The strength at 56 d reached 113 MPa, with the modulus of elasticity 36 GPa.[85]
3. For the columns of the Degeno Office Building in Eschborn, Germany, a 28 d cube strength of 105 MPa was specified for the concrete. The mix proportion for the concrete was as follows: cement content of 510 kg/m^3, silica fume of 35 kg/m^3, water of 138 kg/m^3, water/binder ratio of 0.25. The actual strength obtained was 130 MPa, with a modulus of elasticity of 50.9 GPa.[85]
4. The Pacific Ocean Center No.1 in Seattle, WA in the United States was constructed in 1988, with the column produced from steel tube concrete with a concrete strength of 131 MPa.[85]
5. In the Square Building in Seattle, WA in the United States built in 1988, resistance to earthquakes was considered along with diminishing the swaying and movement due to wind, so the modulus of elasticity for the concrete was enhanced. This building is a 58-story tower, 230.17 m high. The main vertical load-bearing structure consists of four large columns with a diameter of 3 m and 14 columns with diameter of 1.36 m arranged around the four main columns.

The structural system for this building was the structural engineer, Skilling Ward Magnusson Barkshire system (SWMB), characterized by forming a concrete column inside a steel tube, concrete column. In this case, super-high-strength concrete with design strength of 96 MPa was used with the actual 56 d strength reaching 135 MPa and a modulus of elasticity of 50 GPa. The water/binder ratio for the concrete was 0.22 and the slump of the mix 250 mm. The concrete mix was transported and placed by pumping through a trunk pipeline, 122 m long, in continuous casting sections, two stories high (7.3 m). To eliminate the voids and other defects formed during placement, the mix was pumped from the bottom up without vibration into the steel tube mold through an opening in the base. The steel tubes were installed four stories in advance of casting and the construction rate was two stories in 3.5 d. The use of super-high-strength concrete in combination with a specific structural decision lowered the structural cost by 30%. It should be pointed out that this project was completed without a suitable design standard and can be considered an example of the application of super-high-strength concrete in high-rise buildings.[86]

6. In 1983, a building known as La Laurentienne was built in Canada to verify the feasibility of conveying and pouring 120-MPa super-high-strength concrete and an experimental column was produced. The mix proportion for the concrete was cement content of 500 kg/m^3, silica fume of 30 kg/m^3, water/binder ratio of 0.24. The strength of the concrete at 91 d was 120 MPa.[85]
7. In Kuala Lumpur, Malaysia, 3000 m^3 of C80 grade concrete was specified for the profiled steel bar concrete column used in the lower part of the 450-m-high twin towers of Petronas Towers. The binder content of the concrete was Portland cement 260 kg/m^3, Portland fly ash blended cement (fly ash content 19%) 260 kg/m^3, silica fume of 30 kg/m^3, water/binder ratio of 0.27, with superplasticizer added to give a slump reaching 200 mm. The 150-mm cube strength at 28 d reached 100 MPa.[85]
8. In Ableton, Sweden, poles for a power transmission line have been manufactured since 1990 by the centrifuge method to gain a high prestress load and reduce the dead weight as well as to enhance the cracking load, A water/binder ratio of 0.28 was used giving a concrete cube strength of 100 MPa.[85]
9. In Staffanstorp, Sweden, a precast beam was produced from concrete with a water/binder ratio of 0.22 and cube strength of 143 MPa.[85]
10. To solve the problem of excess radial pressure within the tunnel of the Rockenstrasse Potash Mine of Germany in 1988, a super-high-strength concrete segment with a strength of 105 MPa was produced. The materials used were low alkali cement of 450 kg/m^3, silica fume of 45 kg/m^3, superplasticizer of 3%, and a water/binder ratio of 0.17.[85]

11. In 1992, to gain a larger space in an eight-story building containing heavy machines at Bauer Druck in Germany, a precast column made from concrete with a 56 d cube strength of 118 MPa was installed. The materials for the concrete consisted of cement content of 450 kg/m^3, silica fume of 35 kg/m^3, water of 50 kg/m^3, with a water/binder ratio of 0.31.[85]
12. To avoid moisture damage, the relatively low moisture content provided by self-drying of a reinforced concrete slab with a cube strength of 133 MPa and water/binder ratio of 0.27 was produced in Bara of Sweden in 1990 and laid on the ground.[85]
13. On the Ranasfoss Bridge, Norway, an 80-mm layer of steel fiber concrete with cube strength of 110 MPa was laid on the pavement in 1989 to improve the wear resistance of the pavement.[85]
14. The pavement of many expressways in northern European countries has been made from super-high-strength concrete with a compressive strength of 135 MPa to enhance the wear resistance of the pavement. This is because in northern European countries, the studded tires on cars used to add skid resistance increases the wear on the concrete. The abrasion resistance of 135-MPa concrete is more than four times that of 55-MPa traditional concrete.[87]
15. In Barcelona, Spain, the Montjuic Crosswalk Bridge constructed in 1992 was made from concrete with a compressive strength of 100 MPa, to allow a narrow cross section to be used to overcome visual effect restrictions. In 1993, 100-MPa super-high-strength concrete was used in the box-type truss structure of the Takenaka Crosswalk Bridge in Japan to reduce its height.[85]
16. In 1989, 60 m^3 of super-high-strength concrete with cube strength of 140 MPa was used to enhance wear resistance in the Smestad Tunnel in Oslo, Norway, for repairing the slots in the 55-MPa concrete, which had formed after 7 years of operation. The water/binder ratio of the concrete was 0.23 and the old and fresh concretes were bonded together with an epoxy resin.[85]
17. In 1991, to repair the structure of a factory in Israel, subject to the corrosive environment of the Dead Sea, 110-MPa super-high-strength concrete was applied with a water/binder ratio of 0.3, cement content 350 kg/m^3, silica fume 75 kg/m^3, and slump of 160 to 200 mm.[85]
18. High-rise housing (43 stories, 145 m high) in the northern region of Ubasute, Japan, was constructed in 1997. Super-high-strength concrete with standard design strength of 100 MPa was applied in full-scale construction to give a wide living space with a column spacing of 10 m. Formerly, it had been impossible to span 10 m in a reinforced concrete building over 40 stories.[70]
19. The Beijing Finance and Tax Building was completed in 1996 with a building area of 48,000 m^2, three basement stories, and 15 stories

above ground. Concrete with design strength of C60 was used for the building below two stories. In four columns of the first floor, super-high-strength concrete was used as a test to check the construction performance of the concrete mix. The mix proportion was as follows: 525 OPC of 540 kg/m^3, silica fume of 60 kg/m^3 from Tangshan Iron Alloy Factory, a water/binder ratio of 0.23, and 1.5% by weight of DSF-2 superplasticizer. The slump of the mix was 202 mm and 194 mm after 90 min. The 28 d strength (15 cm × 15 cm × 15 cm cube) was 127 MPa, impermeability was more than S35 after 28 d of accelerated carbonation, the depth was zero, the mass loss after 1000 freezing and thawing cycles was zero and the relative modulus of elasticity loss was 2.2%, shrinkage 0.06 mm/m, and creep coefficient was 0.709. The test was successful, this being the first case in China where super-high-strength concrete with such high strength had been used.[88]

20. The Fulin Building in Shenyang completed in 1991 with a building area of 80 thousand m^2 has 2 stories of basements and 32 stories above ground, reaching 125 m in height. A frame structure with a sheer wall was designed by the Liaoning Building Design Institute and steel tube, concrete compound column was used for stories 1 through 6. The design strength of the concrete in the steel tubes was C90 grade, with C60 grade used for the external concrete. In construction, the concrete in the tube was designed for 100 MPa with mix proportions as follows: 42.5R cement of 460 kg/m^3, mineral additive (specific surface 1050 m^2/kg) of 110 kg/m^3, water/binder ratio of 0.26, admixture of 25.7 kg/m^3, coarse sand with a fineness modulus of 3.2 of 770 kg/m^3, 5 to 25 crushed stone (cube compressive strength 143 MPa) of 1020 kg/m^3. The slump of the mix was 240 to 260 mm, with a spread of 600 to 700 mm. The slump loss with time: after staying for 5 h 20 mm, dynamic state (keep mixing in concrete mixer) for 8 h 20 mm, dynamic state for 10 h 40 mm. The average strength of 76 groups of specimens was 115 MPa (converted to the strength of 15 cm × 15 cm × 15 cm cubes by multiplying by a conversion factor of 0.9); the strength test carried out between 60 and 120 d showed the growth of strength. This project illustrates that C100 super-high-strength concrete has been used successfully in China. In addition, in the Dalian Bianye Bank Building (52 stories, 185 m high, building area 230 thousand m^2) C100 super-high-strength concrete in conjunction with application of the steel tube, concrete compound columns was used successfully as well.[89,90]
21. During construction of the City Square of Hong Kong, C100 super-high-strength concrete was used in the design with a cement content of 500 kg/m^3, silica fume replacement of 10%, fly ash replacement of 15%, water/binder ratio of about 0.22, crushed granite aggregate, and addition of superplasticizer and retarder, giving a slump of more than

220 mm, slump loss after 4 h of less than 40 mm, and an average strength of 115 MPa.[27]

22. The National Grand Theater, with a total building area of 145,000 m^2 (1400 m^2 underground and 53,600 m^2 above ground) and a total height of 45.35 m, was designed by a French design company. In this project, C100 super-high-strength concrete was used in some of the columns, 42.5 OPC was used, medium sand (fineness modulus 2.8), crushed stone with crushing index 6.3% and maximum particle size of 25 mm, compound admixture, and superplasticizer. The mix proportion was water/binder ratio of 0.26, cement 450 kg/m^3, additive 150 kg/m^3, sand 614 kg/m^3, crushed stone 1092 kg/m^3, giving a slump of 250 to 260 mm, spread 600 to 620 mm, with a slump after 5 h of 235 to 240 mm. The mass loss after 500 freezing and thawing cycles was 0; the relative dynamic modulus of elasticity loss was 6.9% to 7.6%. The average strength of 21 groups of 150 mm × 150 mm × 150 mm cubes was 117.9 MPa, and a mean square deviation of 6.75. This again demonstrates a successful application of C100 super-high-strength concrete.[91]
23. The Shenyang Yuanji Building (2 basement stories and 28 stories above the ground reaching 96 m in height) is a high-rise building made with steel tube, concrete compound columns designed by the master Lin Liyan from the Liaoning Building Design Institute. The concrete in the steel tubes was specified as C100, with C40 outside the tube. The materials used were 42.5 Portland cement, medium sand, crushed stone (crushing index 6.1%, maximum particle size 25 mm), compound additive (specific surface area 780 m^2/kg) and superplasticizer. The slump of the mix was 250 to 260 mm, with a spread of 600 to 650 mm. After 4 to 4.5 hours, there was little slump loss and the mix poured easily without vibration and self-leveled when, during construction, the steel tube was hammered. The strength grade of the concrete conformed to C110.[92]
24. At some of China's concrete precast plants, elements of super-high-strength concrete with a compressive strength up to 100 MPa have been produced. For example, at the Anhui Bengbu Cement Product Plant, a large number of reinforced concrete supports and guide plates for mine tunnels for the coal industry with strength of 100 MPa have been produced.
25. The Guangzhou Yangcheng Pipe Pile Co. Ltd. mass produces super-high-strength concrete pipe piles with a compressive strength of more than 100 MPa using a technology based on Portland cement + ground sand and autoclave curing.
26. Across the Fulda river in Kassel city, Germany, a 133.2-m-long truss-designed pedestrian bridge with a main span of 36 m was built. The upper chord of the truss is a prestressed ultra-high-strength concrete slab, while the lower chord and the struts are steel tubes, in which

the concrete contains 0.99% of steel fiber by volume giving a mean concrete strength of 165 MPa.[93]

27. During 2003–2004, the Millau Viaduct tollgate in France was constructed. The roof cover, 98 m long and 28 m wide, and composed of 53 precast prestressed concrete members 10 cm thick, was made from ultra-high-strength concrete. Materials used per m^3 concrete were superplasticizer 44.6 kg, water content of 195 kg (W/C = 0.22), and a steel fiber content of 195 kg. The mean 28 d strength of the concrete was 199 MPa.[94]
28. In the Drome region of France, a two-span highway bridge was built from ultra-high-strength concrete in 2001. Each span was 22 m, the cross section composed of five Π-shaped precast prestressed beams, 2.4 m wide, 20.5 m long, and 0.9 m high with the web plate 11 cm thick. The weight of each beam was 37 t, with an apparent density of 2800 kg/m^3. The 28 d strength of the concrete was up to 175 MPa, and the modulus of elasticity of 64 GPa. The materials used per m^3 were cement 1114 kg, silica fume 169 kg, 0 to 6 mm aggregate 1072 kg, steel fiber 231 kg, superplasticizer 40 kg, water 209 kg, giving a spread of 640 mm, without vibration.[95]

In addition, super-high-strength concrete has been used in the cooling towers of nuclear power stations.[94]

11.2 THE ECONOMY OF APPLICATION FOR SUPER-HIGH-STRENGTH HIGH PERFORMANCE CONCRETE

According to Gao Xueshan,[96] a prestressed railway bridge with spans of 50 m and 80 m, constructed with ordinary, high strength and super-high-strength concretes, will lead to large differences in the height of the beam, concrete volume, and weight of the beam. These indexes will be decreased greatly with an increase in the compressive strength of concrete (Table 11.1).

It can be seen from Table 11.1 that in the case of the 50-m span made from 120-MPa super-high-strength concrete, the beam height was reduced from 3.1 m with 40-MPa ordinary concrete to 1.7 m, a reduction of 45.2%. The weight of the beam was reduced from 165 t to 90 t, a reduction of 44.8%. In the case of the 80-m span made from 120-MPa super-high-strength concrete, the beam height was reduced from 5.8 m for 40-MPa ordinary concrete to 2.8 m, a reduction of 55.2%, and the weight of the beam was reduced from 545 t to 245 t, a reduction of 55%. This clearly shows the great technical and economical benefits from the application of super-high-strength concrete.

Because the compressive strength of super-high-strength concrete is very high, a suitable area for its application is acting as compression-bearing

Table 11.1 Comparison of railway bridges using ordinary concrete, high strength concrete, and super-high-strength concretes

Type of concrete	*Compressive strength (MPa)*	*Span (m)*	*Height of the beam (m)*	*Height/span ratio*	*Concrete volume (4 pieces) (m³)*	*Mass of each beam (t)*
OC	40		3.1	1:16	262	165
HSC	80	50	2.1	1:24	173	110
SHSC	120		1.7	1:30	145	90
OC	40		5.8	1:14	872	545
HSC	80	80	3.6	1:22	506	315
SHSC	120		2.8	1:29	391	245

Note: OC = ordinary concrete; HSC = high strength concrete; SHSC = super-high-strength concrete.

elements in civil engineering such as in the columns in high-rise buildings as well as the pylons and arch rings in bridges. Taking the column as an example, in Nawy and Eng[4] and Webb,[97] there are three kinds of reinforced concrete columns introduced, namely, one kind with a high strength concrete column with concrete strength of 60 MPa; two kinds with 120-MPa super-high-strength concrete columns; two kinds with steel tube concrete with concrete strength of 60 MPa and 120 MPa for each; and, in addition, one kind of I-shaped steel column (Table 11.2). These six column types all have the same bearing capacity. Taking the cost of a square reinforced concrete column with a cross section of 840 mm × 840 mm and 60-MPa concrete as 1, the cost of a column of steel tube containing super-high-strength concrete with diameter of 570 mm and thickness of 8 mm and concrete strength of 120 MPa is 0.71, a reduction of 29%. The cost of a square reinforced concrete column with a cross section of 600 mm × 600 mm and a concrete strength of 120 MPa is 0.79, a reduction of 21%; the cost of a square reinforced concrete column with cross section of 660 mm × 660 mm and concrete strength 120 MPa is 0.77, a reduction of 23%; the cost of a steel tube, high strength concrete column with diameter of 740 mm and steel tube wall thickness of 8 mm, and a concrete strength of 60 MPa is 0.98, a reduction by 2%; the cost of a steel column with flange dimensions of 600 mm × 40 mm and web plate dimensions 520 mm × 40 mm (steel grade 350) is 2.21, an increase of 121%. It can be seen from Table 11.2, which compares the costs, that for a reinforced concrete column or steel tube, concrete column, and for the column with 120-MPa super-high-strength concrete, they all have a much lower cost in comparison with the column from 60-MPa high strength concrete, without considering the economic benefits from enhancement of the usable floor area due to the reduction of the cross section of the column. It shows that using super-high-strength concrete is reasonably economical.

Table 11.2 Relative expenses for the columns of different kinds

No.	*Type of column*	*Parameters for the column*	*Relative expense*
1	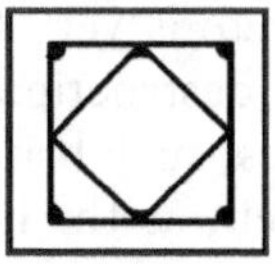	Cross section 840 mm × 840 mm (33 in × 33 in) 6Y 24 mm (6–1 in steel bars) AS 3600 stirrup (R10–360) Concrete strength 60 MPa	1.0
2	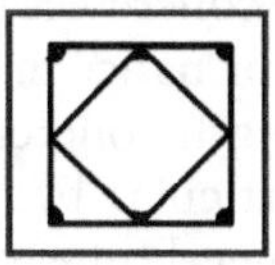	Cross section 600 mm × 600 mm (24 in × 24 in) 8Y 24 mm (8–1 in steel bars) crawley stirrup (Y32–200) Concrete strength 120 MPa	0.79
3	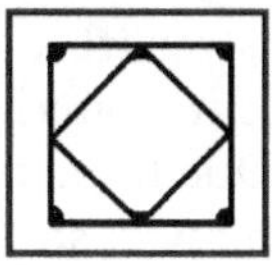	Cross section 660 mm × 660 mm (26 in × 26 in) 8Y 24 mm (8–1 in steel bars) AS 3600 stirrup (R10–360) Concrete strength 120 MPa	0.77
4	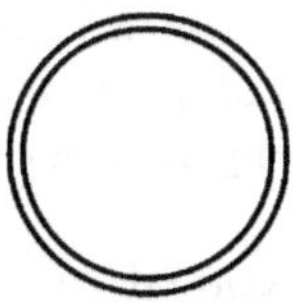	Steel grade 250 steel tube 740 mm × 8 mm (29 in diameter, 3/8 in thick) Concrete strength 60 MPa	0.98
5	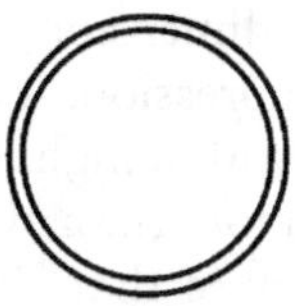	Steel grade 250 steel tube 570 mm × 8 mm (23 in diameter, 5/16 in thick) Concrete strength 120 MPa	0.71
6		Steel grade 350 steel column Flange of column 600 mm × 40 mm Web plate 520 mm × 40 mm	2.21

Note: 1 in = 2.54 cm.

11.3 THE BEST APPROACHES FOR APPLICATION OF SUPER-HIGH-STRENGTH HIGH PERFORMANCE CONCRETE—KILOMETER COMPRESSIVE MATERIAL

Japaneses researchers have carried out a laboratory study on the application of super-high-strength concrete with a strength grade more than 150 MPa in a reinforced concrete column (there have been no practical applications up to now).[70] However, the author thinks that it is not reasonable to use ordinary reinforcement allocation for super-high-strength concrete. Concrete is well known to be a brittle material. Ordinary concrete has a low brittleness, and as we can see from the stress–strain curve, internal microcracks develop, and certain plastic deformations occur (pseudo-plastic deformation). With an increase in strength, plastic deformation occurs less and less, and the brittleness increases. Super-high-strength concrete is very brittle, its stress–strain curve is a straight line, so that during its destruction, catastrophic damage occurs without any forewarning. Besides, such brittleness cannot be overcome by super-high-strength concrete itself, as the bond in the microstructure of the aggregate or cement paste of the concrete is mainly covalent. Super-high-strength concrete can be a low performance material. Therefore, the most suitable way to use super-high-strength concrete is in an effective combination with a metallic material, in particular by the all-sided restriction of super-high-strength concrete with a metal tube to make a composite material, namely, a steel tube, ultra -high-strength concrete for use. Therefore, the author has studied such a composite material, and a steel tube, ultra-high-strength concrete was prepared. Due to its tremendous bearing capacity and excellent deformation performance, steel tube, ultra-high-strength concrete can be used for construction of a kilometer-high, super skyscraper, so the author calls it *kilometer compressive material.* Following is the discussion on how super-high-strength concrete can be applied to make kilometer compressive material.

The term *kilometer compressive material* has been coined to indicate material that has an engineering acceptable section and a dimension extending to over 1 km in the direction of a member under compression. It can be used as the bearing column of a super skyscraper over 1000 m high, for the arch ring of a super-large-span arch bridge with a span more than 1000 m, or the bearing column or tube of a marine structure more than 1000 m deep. One of the differences from other loading states (tension, bending, shear, and torsion) is that when a column is under compression, longitudinal buckling can be an issue; the longer the column, the larger the longitudinal buckling and the less its bearing capacity. To reduce the influence of any longitudinal buckling in the structure, lateral supports or partitions must be allocated to reduce the calculated length of the bearing column and enhance its stability. Even when scaling takes into account the increased height of the structure, the load on the material at the bottom will be high so that the strength of the material may not meet the load requirements and a material with greater strength and bearing capacity is needed.

Thus, the requirements for a kilometer compressive material for use in the construction of super structures such as a kilometer-high, super skyscraper are as follows:

1. Extra-high strength and large bearing capacity of structural members.
2. High modulus of elasticity so the kilometer compressive material will have little deformation under high loading.

3. Good ductility. Highly ductile material will not suffer from brittle failure under dynamic loads caused by hurricanes, waves, and earthquakes.
4. Relatively low dead weight so the total mass of the building can be reduced.
5. Excellent durability. A building using such material can stand for a long time.
6. Abundant resources for raw materials. Due to the very large volume of a super structure, large amounts of materials will be consumed. Therefore, there should be an extensive supply of raw materials available as well as a simple technology for manufacture together with low cost.

At present, three kinds of materials are widely used by mankind, namely, organic, metallic, and inorganic non-metallic materials with mainly steel and concrete of various strength grades used for structural applications. Skyscrapers with more than 100 stories, such as the Sears Tower in the United States, the World Trade Center Tower and the Empire State Building in New York City, and so forth, were made with steel as their main structural material. But the Water Tower Square (94 stories) in Chicago, the Ryugyong Hotel (105 stories) in Korea, the Metropolitan International Building (104 stories) in Bangkok, the Central Plaza (70 stories) in Hong Kong, Jinmao Tower in Shanghai, and the second tallest building in the world, the Petronas Tower in Kuala Lumpur, have concrete as their structural material.

Steel has high strength and good ductility and a super-high-rise building constructed from steel has low weight but also insufficient stiffness.

A reinforced concrete super-high-rise building has a heavy dead weight, but has high stiffness and can be constructed at a relatively low cost. By enhancing the concrete strength, its bearing capacity will be greatly increased and the dead weight lowered. The success of manufacturing super-high-strength concrete with a compressive strength of 100 to 150 MPa[35] will make an important contribution to lowering the dead weight and increasing the bearing capacity of the concrete structure. However, as described above, with its high brittleness, using ordinary reinforcement allocation in super-high-strength concrete is not appropiate.

A new type of material, carbon nanotubes with a strength 100 times greater than steel and Young's modulus as high as 1 TPa (10^{12} Pa) and with several percent elongation,[98] has a mass much lighter than steel. Theoretically, such a new nanometer material is a strong candidate for the kilometer compressive material. However, currently such material is only produced as a laboratory material and it is very expensive, several times that of gold. At present, the practical application for such a material for kilometer compressive material is impossible.

Active powder concrete may reach 200 to 800 MPa in strength. However, at present this super-high-strength material cannot be used for kilometer compressive material as it is produced by a special *hot pressing* technology and the raw material is refined (by removing the coarse aggregate) leading to a great increase in cost.

Therefore, considering the resources available for mass building structural materials at present and the experience accumulated in engineering, only steel and concrete can be taken as the basis for preparing kilometer compressive material. However, both have disadvantages and cannot fully meet the six requirements given above and it is not reasonable to use them separately. Consequently, it is necessary to combine both these materials to overcome their disadvantages so that they can be used to their best advantage. Second, the strength for both materials should be increased. So, the technical approach to obtain a kilometer compressive material should use super-high-strength steel and super-high-strength concrete, effectively combining two super-high-strength materials.

During loading, the in-filled concrete is restricted by the steel tube and subject to 3D compression. This combination is an effective way to overcome the disadvantages of both steel and concrete and give full play to their advantages. In this way, super-high-strength steel tube, concrete composite material may become the most practical, available kilometer compressive material meeting the six requirements given above. This is because

1. The problem of the high brittleness of super-high-strength concrete can be solved completely. The super-high-strength concrete within a steel tube is restricted under compression, and approaching rupture, will not undergo a sudden breakdown, the whole composite material showing large ductility with the start of plastic deformation of the steel. This effective integration allows the highly brittle super-high-strength concrete to be transformed into a ductile super-high-strength steel tube, concrete composite material, overcoming the large brittleness of the super-high-strength concrete. Our experimental study[14,99] has proved this.
2. Steel has high strength and good ductility, but a steel tube has a thin wall and under high loading stress, damage may occur due to the instability of the tube wall. After filling the tube with concrete, the concrete core will give strong support to the steel tube enhancing the stability of the tube wall. In such an integration of a steel tube with super-high-strength concrete, both materials offset the weak points of each other with their own strong ones to provide perfect integration.
3. Both the super-high-strength steel tube and the super-high-strength concrete have high strengths, and after integration of the super-high-strength concrete in the steel tube the concrete is in 3D compression due to the restriction on lateral deformation, so the strength is enhanced

greatly compared with 1D compression. During compression, the steel tube is transformed gradually from axial loading into an annular tension. At this point, the concrete mainly bears the axial compressive stress while the steel tube carries a tensile stress, which restricts lateral expansion of concrete, giving full play to the advantage of the steel under a tensile load. Thus, the composite material will have a greater bearing capacity than the sum of the bearing capacities of steel tube and concrete, in effect making 1 + 1 > 2.

In addition, if the strength of the concrete is increased from the 30 MPa for ordinary concrete to 150 MPa for super-high-strength concrete, the unit mass is only increased from 2400 kg/m^3 to 2600 kg/m^3, which is a small increment.

It can be seen from the above information that from the point of view of strength, stiffness, ductility, dead weight, and raw material resources that a super-high-strength tube/concrete composite material is an ideal candidate for the kilometer compressive material.

To produce this kilometer compressive material in reality, the author has, in the Laboratory of Concrete Technology at Chongqing University, prepared 29 short steel tube, ultra-high-strength concrete columns with a hoop index of 0.415 to 1.249. The ultra-high-strength concrete had a compressive strength of more than 150 MPa, with a 90 d compressive strength of 156.7 MPa to 164.9 MPa, an axial compressive strength of 134.5 MPa and 141.0 MPa, and in a φ138- to 140-mm steel tube with a yield strength, f_a, of 313 to 345 MPa. The bearing capacity and deformation properties of the columns have been studied (Table 11.3, Figure 11.1, and Figure 11.2). The coarse aggregate for the concrete was crushed limestone with a maximum particle size of 20 mm, with medium sand used with 625 Portland cement, silica fume, and superplasticizer. The concrete mix had water/binder ratios of 0.20 and 0.18, giving slumps of 258 mm and 228 mm and spreads of 629 mm and 438 mm, respectively, with the time for an upended cone flow of 15 s and 24 s, respectively. The concrete could be placed with existing construction methods.[14] Experiment results show that

1. Ultra-high-strength concrete has a very high brittleness, which under compression ruptures explosively. By restricting the concrete with a steel tube, it is converted into a steel tube, ultra-high-strength concrete composite material with excellent plasticity. The test was terminated at deformation of the specimens, 7.87% to 20.77%. Though there was a large deformation, the specimens did not disintegrate or collapse. At termination of the test, the bearing capacity of the specimen reached 75.3% to 135.2% of the ultimate load, which depended on the hoop index.

Table 11.3 Test results of steel tube, ultra-high-strength concrete short column under axial compression

No.	Specimen mark	$D \times t \times L$ (mm × mm × mm)	A_a (mm²)	ρ (%)	D/t	f_a (MPa)	A_c (mm²)	f_c (MPa)	θ	Ultimate load (kN)	Ultimate deformation (%)	Valley load (kN)	Valley deformation (%)	End load (kN)	End deformation (%)
1	GZG1-2	138 × 5 × 483	2098	14.34	27.6	343	12859	134.5	0.416	2490	0.80	2090	1.66	2320	13.48
2	GZG1-3	138 × 5 × 483								2430	0.96	1620	2.02	1830	7.87
3	GZG2-1	138 × 7 × 483	2881	19.26	19.7	343	12076	134.5	0.608	2750	0.93	2340	3.21	2580	14.35
4	GZG2-2	138 × 7 × 483								2870	1.34	2550	1.84	2660	11.89
5	GZG2-3	138 × 7 × 483								2880	1.84	2440	3.15	2605	11.41
6	GZG3-1	140 × 9 × 490	3704	24.06	15.6	345	11690	134.5	0.813	3160	1.07	2880	3.17	3400	10.88
7	GZG3-2	140 × 9 × 490								3170	1.35	2800	2.19	3650	12.92
8	GZG3-3	140 × 9 × 490								3240	0.99	2690	1.86	2700	13.14
9	GZG4-1	140 × 11 × 490	4458	28.96	12.7	343	10936	134.5	1.046	3340	0.99	3180	1.62	3930	13.87
10	GZG4-2	140 × 11 × 490								3310	1.29	3200	1.83	3950	13.40
11	GZG4-3	140 × 11 × 490								3350	078	3080	1.21	3850	12.03
12	GZG5-1	140 × 12.5 × 490	5007	32.52	11.2	313	10387	134.5	1.122	3410	0.95	3250	1.34	4400	17.90
13	GZG5-2	140 × 12 × 490	4825	31.10	11.7	345	10569	134.5	1.171	3400	1.21	3350	2.30	4300	15.61
14	GZG5-3	140 × 12 × 490								3420	1.31	3340	2.00	4300	13.76

15	GZG6-1	138 × 5.7 × 483	2369	15.84	24.2	343	12588	141.0	0.458	2780	0.84	2050	1.89	2530	14.16
16	GZG6-2	138 × 5.7 × 483								2700	0.68	1840	2.38	2550	14.93
17	GZG6-3	138 × 5.7 × 483								2705	0.89	2180	2.84	2700	16.38
18	GZG7-1	140 × 8 × 490	3318	21.55	17.5	345	12076	141.0	0.672	3200	0.90	2660	1.78	3400	15.50
19	GZG7-2	140 × 8 × 490								3170	0.78	2770	2.04	3500	14.42
20	GZG7-3	140 × 8 × 490								3130	0.59	2620	1.35	3500	16.21
21	GZG8-1	140 × 10 × 490	4084	26.53	14.0	313	11310	141.0	0.802	3130	0.62	2960	1.23	3720	16.51
22	GZG8-2	140 × 10 × 490								3050	1.47	2926	1.62	4000	20.77
23	GZG8-3	140 × 10 × 490								3200	0.76	2940	1.13	3750	15.39
24	GZG9-1	140 × 12 × 490	4825	31.35	11.7	313	10568	141.0	1.014	3280	0.87	3270	1.13	4200	15.36
25	GZG9-2	140 × 12 × 490								3240	1.06	3170	1.63	4050	16.82
26	GZG9-3	140 × 12 × 490								3270	0.65	3050	0.86	4040	14.94
27	GZG10-1	140 × 14 × 490	5542	36.00	10.0	313	9852	141.0	1.249	3550	1.44	3510	1.62	4800	16.85
28	GZG10-2	140 × 14 × 490								3500	0.85	3450	0.98	4800	18.76
29	GZG10-3	140 × 14 × 490								3410	0.58	3320	0.74	4800	16.93

Note: $D \times t \times L$ = diameter × thickness × length; A_a = area of steel tube; ρ = steel content; D/t = diameter thickness ratio; f_a = yield strength of the steel; A_c = area of the concrete; f_c = axial compressive strength of the concrete; θ = hoop index.

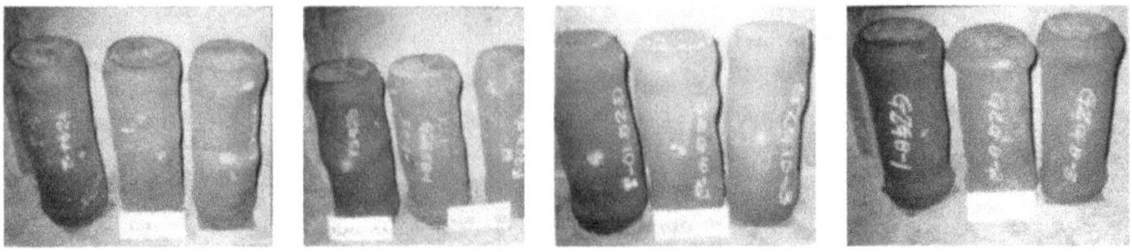

GZG4 GZG5 GZG8 GZG10

Figure 11.1 Appearance of specimens GZG4, GZG5, GZG8, and GZG10 after testing.

In addition, in spite of the diameter/thickness ratios of most of steel tubes used in the test being less than 20 (CECS 104:99 specifies that the diameter/thickness ratio should be limited in the range of $90(235/fa) - 20$), the steel tube, but thick wall steel tube did not show local yielding.

2. Using the equation for bearing capacity of steel tube, high strength concrete suggested by Cai Shaohuai,[100] where he suggested

$$N_c = A_c f_c (1 + K\theta)$$

and based on test results from 29 specimens where K = 1.21,[14] the 1.20 used for K is reasonable. Therefore, the bearing capacity of a steel tube, ultra-high-strength concrete short column can be expressed as follows:

$$N_c = A_c f_c (1 + 1.2\theta) \tag{11.1}$$

where

N_c = bearing capacity of the steel tube, ultra-high-strength concrete short column, N_0;
A_a = cross section of concrete core, mm^2;
f_c = axial compressive strength of ultra-high-strength concrete, MPa;
θ = hoop index,

$$\theta = \frac{A_a f_a}{A_c f_c}$$

where

A_a = cross section area of the steel in mm^2;
f_a = yield strength of the steel, MPa.

With an enhancement of the concrete strength grade, the lateral deformation of the concrete decreased with the hoop effect of the steel on the concrete also decreasing, as shown by the k value being lowered. According to the

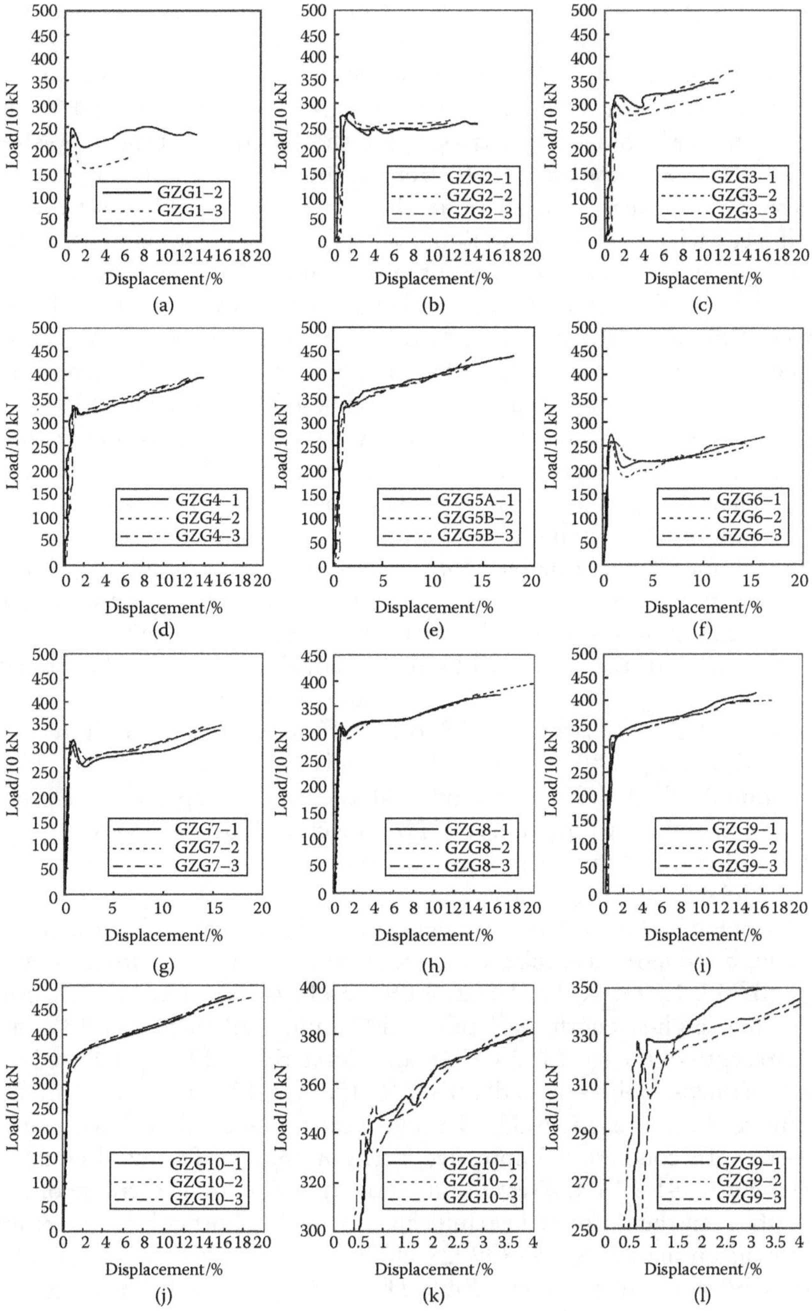

Figure 11.2 Load-displacement curves of the steel tube ultra-high-strength concrete short column.

Technical Regulation of High Strength Concrete Structures, CECS 104:99, for a steel tube concrete containing high strength concrete, the k value decreased from 2.0 for C50 with a rate of 0.05 to 1.70 for the C80. The strength of concrete used in our test reached as high as 156.7 to 164.9 MPa, so lowering the k value to 1.21 is logical. It showed also that the hoop effect of the steel tube on the ultra-high-strength concrete is significant.

Despite the decrease of the k value, the f_c value was very high, making the bearing capacity, calculated by Equation (11.1), large. For example, taking the acceptable external diameter of the steel tube, ultra-high-strength concrete as D = 2500 mm (the external diameter of the largest steel tube, concrete column in the world, at Two Union Square in Seattle, WA in the United States, is 3000 mm), $f_a = 345$ MPa, $f_c = 140$ MPa, $\theta = 0.72$, and $A_c =$ 3,801,336 mm^2, the bearing capacity calculated by Equation (11.1) $N_c =$ 9.92×10^5 kN is 99.2 thousand tonne. If we take as an analogue, the tallest steel tube, concrete super-high-rise building in the world, the Saige Building in Shenzhen, China built in 1997, which has 76 stories (72 stories above ground) and is 291.6 m high, the maximum axial load for a single column is 9000 tonne.[101] It can be seen that if steel tube, ultra-high-strength concrete with its large bearing capacity was used, a kilometer super skyscraper with more than 300 stories and over 1000 m high is possible.[110]

The strength of the steel used by the author did not match the strength of the concrete, $f_a/f_c = 2.2$ to 2.6 in the built steel tube, concrete high-rise buildings in China, where $f_a/f_c = 5.9$ to 11.8. If super-high-strength steel had been used, where $f_a = 800$, 1000, or 1300, and given $f_c = 140$ MPa, D = 2500, 3000 mm, $\theta = 0.75$, 1.00, 1.25, and 1.50, the bearing capacity of the short column calculated by Equation (11.1) will reach 110 to 240 thousand t.[102,111] There will come a day when construction engineers will build a column with such large bearing capacity.

Subsequently, the author's student, Dr. Wang Yongwei, produced a super-high-strength steel tube, concrete short column with a different hoop index (0.945, 1.226, 1.512, 1.802, 2.097) and f_a/f_c enhanced to 7.27, using S135 super-high-strength steel tube (ϕ127 mm, wall thickness 9.19 mm, yield strength of the steel 1026 MPa, and elongation 9.29%) processed into tube specimens with various diameter (ϕ118, 120, 122, 124, 126 mm) and thickness (3.5, 4.5, 5.5, 6.5, 7.5 mm), which was combined with ultra-high-strength concrete (compressive strength 163.7 MPa, axial compressive strength 141.2 MPa). Test results show that due to the increase of steel strength, even the ultimate bearing capacity of the short column increased greatly (attaining 2724, 3448, 3921, 4230, 4599 kN, respectively) and its deformability is still excellent (Table 11.4 and Figure 11.3), showing good prospects for kilometer load-bearing material with excellent deformability and high bearing capacity produced by super-high-strength steel and ultra-high-strength concrete.[103]

Table 11.4 Test results for short columns with super-high-strength steel tube and ultra-high-strength concrete under compression

No.	Specimen mark	$D \times t \times L$ (mm × mm × mm)	ρ (%)	D/t	f_a (MPa)	f_c (Mpa)	θ	Ultimate load (kN)	Ultimate deformation (%)	Valley load (kN)	Valley deformation (%)	End load (kN)	End deformation (%)
1	NSCC1-1	118 × 3.5 × 413	11.5	33.7	1026	141.2	0.945	2792	1.17	2390	5.23	2640	6.47
2	NSCC1-2							2680	0.83	2140	3.39	2740	7.13
3	NSCC1-3							2700	0.98	2290	3.19	2450	5.00
4	NSCC2-1	120 × 4.5 × 420	14.44	26.7	1026	141.2	1.226	3410	1.22	2962	2.98	3010	4.61
5	NSCC2-2							3460	1.06	3128	5.69	3180	7.55
6	NSCC2-3							3475	1.00	2980	3.43	3250	6.38
7	NSCC3-1	122 × 5.5 × 427	17.22	22.2	1026	141.2	1.512	3950	1.24	3435	4.79	3688	7.32
8	NSCC3-2							3942	1.14	3624	5.44	3710	7.19
9	NSCC3-3							3870	1.18	3435	5.21	3620	7.30
10	NSCC4-1	124 × 6.5 × 434	19.87	19.1	1026	141.2	1.802	4092	1.15	3920	2.08	3960	3.49
11	NSCC4-2							4260	1.31	4075	2.87	4100	4.15
12	NSCC4-3							4338	1.36	4088	3.79	4120	5.59
13	NSCC5-1	126 × 7.5 × 441	23.39	16.8	1026	141.2	2.097	4531	1.28	4519	1.81	4430	8.18
14	NSCC5-2							4597	1.28	4588	1.70	4350	8.93
15	NSCC5-3							4670	1.29	4528	1.84	4548	4.98

Note: $D \times t \times L$ = diameter × thickness × length; ρ = steel content; D/t = diameter thickness ratio; f_a = yield strength of the steel; f_c = axial compressive strength of the concrete; θ = hoop index.

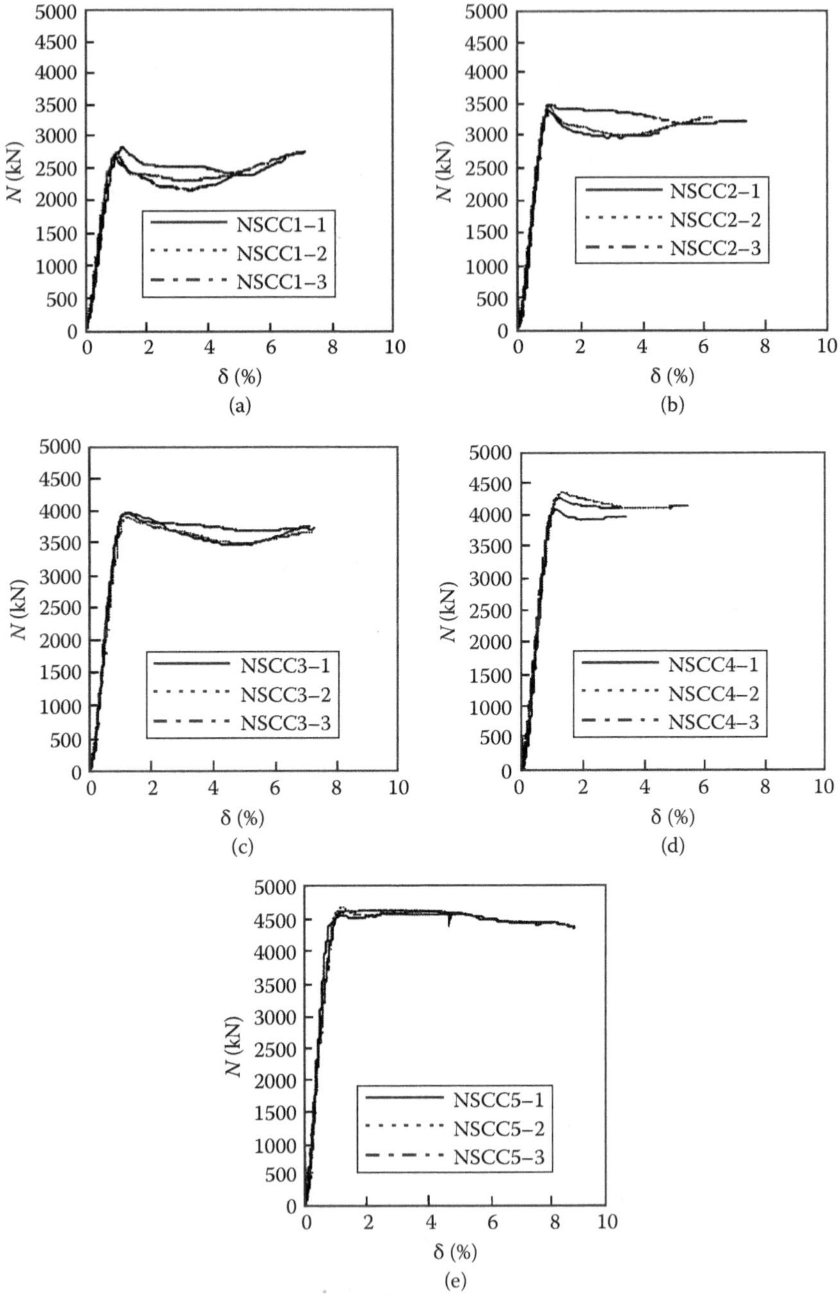

Figure 11.3 Load-displacement curves of a short column with super-high-strength steel tube and ultra-high-strength concrete.

11.4 PROSPECTS FOR SUPER-HIGH-STRENGTH HIGH PERFORMANCE CONCRETE APPLICATIONS AND THE POTENTIAL FOR KILOMETER COMPRESSIVE MATERIAL

11.4.1 Construction of super-high-rise buildings, skyscrapers, and super skyscrapers

Presently, the population of the world exceeds 6 billion, with close to 1.3 billion in China. It is estimated that by 2050, the population of the world will reach 8 billion with 1.5 billion in China. Due to this growth in population, the cultivated land per capita will be greatly reduced. To solve residential and social problems, a great deal of new housing, different transport facilities, and medical and sanitary facilities, industrial bases, commercial facilities, entertainment, and recreational facilities will make massive demands on land resources. Therefore, competition between people and land use is extremely acute; land resources become more valuable year by year. Living space and even manufacturing sites will reach toward the sky or extend underground with spreading over the sea, which is one solution for the problem.

In fact, as building technology has developed, construction of taller and taller buildings has occurred. For example, in 1870, the Equitable Life Insurance Building was built in New York City with elevators in a seven-story building about 25 m high. In 1997, the Petronas Towers were built in Kuala Lumpur, Malaysia to a height of 450 m. Then in 2003, the Taipei International Financial Center was built to a height of 508 m and the Burj Khalifa in Dubai, United Arab Emirates at 828 m, completed in 2010, is the tallest building. In the 1950s, the highest building in China was the Shanghai Park Hotel (24 stories) but nowadays many buildings over 30 stories can be found in all large- and medium-sized cities of China. In 1988, the Jinmao Tower with 88 stories and 420 m was built and has since been followed by the Shanghai World Financial Center at 492 m.

The skyscraper is not only needed for development in the center of crowded cities, but has become the symbol of urban development, demonstrating the progress of science and technology, and economical capability. Therefore, it can be expected that the record of 828 m will be surpassed, and a super skyscraper higher than 1000 m, will emerge. In fact, plans and ideas have already been expressed. For example, an association of 97 Japanese companies organized a study of *super skyscrapers* which finalized a super building design, *Sky City Tower—1000*, 1000 m high, a total area of 1000 ha and a service life of 1000 years. The Japanese Taisei Construction Co. has suggested X-SEED 4000, an imaginative design for a large 3D super sky compound city with a height of 4000 m and a diameter for the foundation 6000 m. The planned Hongkong Bionics Tower will have a height of 1180 m.

There are two key problems in the construction of the skyscrapers mentioned above: one is the selection and development of new structural materials, and the other is the selection of a suitable structural scheme and system. Development and use of a new kilometer compressive material with excellent performance and large bearing capacity could make a reality the imaginative schemes described above.

11.4.2 Construction of large span and super-large-span bridges

In China, the only super-large-span bridges with spans over 1000 m at present are the Jiangying Changjiang Bridge and Hong Kong Tsinma Bridge, but these are cable suspended. In 1997, a reinforced concrete arch deck bridge made with steel tube, high strength concrete as the composite truss was built in Wanzhou, Chongqing. With a span of 420 m, it is among the largest concrete arch bridges in the world. The strength grade of the concrete was C60, and the bridge spanned the river in a single span without a pier in the river.

The construction technology for this bridge was a specific one for such a large span and construction work was carried out without a supporting bracket. In the first step, the steel tube trusses (φ402 mm × 16 mm, with five trusses for each of the upper and lower chords) were lifted by a cable crane across the river, the normal construction method for suspending the assembly of the main beam of a cable-stayed bridge. Thirty-six sections of steel tube truss were assembled one by one from the pier to the crown of the arch on both sides until the crown was closed. In the second step, a C60 concrete mix was pumped simultaneously from both ends of the hollow trusses to form a steel tube, high strength concrete composite truss. The third step used the steel tube, high strength concrete truss and formwork was installed for pouring the outer covered C60 high strength concrete, finally forming a reinforced concrete arch ring with a cross section of three chambers covering the steel tube, concrete truss. The completion of the construction of this bridge marked progress in bridge construction technology.[101] Advanced bridge construction technology used in combination with kilometer compressive material could allow an increase in the span of such bridges.

In China, like the rest of the world, as the population increases and becomes more mobile, rivers, mountains and valleys, thousands of islands, and many straits must be bridged for highways and railways. The growing trend in bridge construction is to build bridges with large and super-large spans that cross rivers, straits, or valleys with a single span. Kilometer compressive material will play an important role in construction of the super bridge with a span over 1000 m.

A number of plans and concepts to connect islands and span straights with super-large-span bridges have been proposed. For example, a bridge connecting the island of Sicily with the Italian mainland has been proposed with a single span, cable-suspended bridge with a main span of 3300 m. An intercontinental bridge across the Strait of Gibraltar, connecting Europe and Africa, has been contemplated. The width of the eastern side of the strait is 14 km with a maximum water depth of 950 m; the width of the western strait is 26 km with a maximum water depth of 300 m. Several multi-span bridge designs have been considered, where the largest span is 5000 m[104]—a cable-suspended bridge, a hanging arch bridge, or a compound suspended, cable-stayed bridge—although the pier and tower will be very high. Kilometer compressive material manufactured from super-high-strength concrete will be very effective in these bridges.

11.4.3 Construction of deep marine constructions and seabed constructions

The oceans and seas make up about three-fourths of the Earth's surface, offering further space for mankind to expand. To explore the natural resources in the oceans and sea (such as oil, natural gas, methane hydrate, and other minerals) many countries have constructed wells and platforms on the seafloor. It can be expected that these marine constructions will expand to deeper water with a depth of 1000 to 3000 m. In this case, kilometer compressive material will be key for exploiting these deep water resources.

Construction on the seabed can be imagined at depths up to 200 m to provide accommodation and work environments. These constructions will bear large water pressure, in which case kilometer compressive material will find its application.

11.4.4 High-rise construction

Mineral fuels such as coal, oil, and natural gas will soon be exhausted and hydropower will be completely developed. Solar energy is endless, but only one 2.2 billionth of its energy reaches the Earth. That energy per second is equivalent to 6 million tonne of standard coal. However, most of the solar radiation is deflected by clouds and atmosphere. In the future, it may be possible to build a solar energy collector and power station in orbit, allowing solar energy to be used effectively. In this case, the problem may be solved by using kilometer compressive material.

Electricity generation using wind power is a new form of power generation because the turbine is driven by wind so no fuel is needed and there is no pollution for the environment. Such energy is totally clean, with a potential capacity of 200 thousand kW. For such power generation, construction of a high tower (similar to a chimney) is needed, preferably in the desert.

Kilometer compressive material is a good prospect for the construction of a wind power station tower.

Kilometer compressive material will find wide application in the construction of television towers, sightseeing towers for tourists, communication facilities, and other high-rise installations to be built.

11.4.5 Construction of artificial green peaks and peak forests

In prehistoric times, the land surface of the Earth was covered in dense forests, with animals running wild. Man initially used caves to live in to avoid carnivorous animals. Later, he built shelters by cutting down trees in which to live while hunting. Then, he destroyed the forest to make cultivatable land as farming increased. With the growth of inhabitants, the amount of cultivated land increased and forests decreased. Throughout the long history of development and evolution of primitive, slave, and feudal societies, the forest was more seriously damaged as a great deal of timber was cut to build palaces for kings and emperors, houses for officials, and luxurious buildings for merchants. The common people also built their housing and temples with timber. In an ancient Chinese descriptive prose, it was said that when the Efang Palace was built, the Mountain Shu became *bald*, a typical description of the forest destruction in China's history. If this pattern goes on, deforestation will continue, so for some regions there will be no trees and the land could turn to desert as top soil is lost, and the rivers could become muddy and silt up so flooding occurs frequently. It may be that the 5000 years of China's history is one of tree cutting. The Loulan Country on the Silk Road, once prosperous, has disappeared due to deforestation. Only today do people gradually understand the importance of the silent forest for the continued existence of mankind. However, it may be too late. On the other hand, as modern industry develops, construction of traffic facilities and extensions of the cities occupy a large area of land. A great deal of mineral fuel has been mined and burnt, yielding a large amount of CO_2 and other toxic gases making cities full of smoke and dust, and the atmosphere dirty and stuffy. In a highly populated city, the ecological environment is worsening. To escape from the pollution in the city, rich people go to the countryside to build country homes. Most people remain in the city with its pollution. Though gardens and parks with planted trees have been built in the cities, these are utterly inadequate measures, and not the sound strategy needed to solve the problem. When the city is mostly occupied by reinforced concrete high-rise buildings and covered by roads, where should trees be planted? However, in today's modern city with its industry, transport facilities, businesses, modern science, technology, and education creating material and spiritual wealth, it is necessary for society to develop further. How should both urban development and planting proceed simultaneously?

The author suggests that to solve this problem, we should transform the white, gray, or other colored urban buildings (except for some historic places) into green buildings covered with plants, using the buildings as carriers for planting. Similarly, roads, bridges (except for traffic lanes), dams, and so forth should be used for tree planting so urban buildings and forest can be merged together.

As an example, in high-rise buildings the roof could be designed as a planting area for vegetation, with the large facade and wall surfaces (except for windows and doors) covered with plants. This can be done easily, with a trough of reasonable width and depth and bottom drainage at the periphery of each floor set up. These troughs can be filled with soil, installed with railings outside and water collection tubes and trees or flowers planted, making a green covering for the building. Thus, the whole building changes into an artificial green tower with a group of high-rise buildings becoming green tower forests. Living and working in such a forest, people can breathe the fresh air emitted by trees and enjoy green and luxuriant forest. When in harmony with nature, one does not lose the convenience and comfort of modern life.

In fact, such a natural forest exists in nature at China's world heritage site, Wulinyuan of Zhangjiajie, where there is an area of quartz peaks covered with forest plants. There are more than 3000 green peaks with more than 200 over 1000 m high, which is one of the famous tourist spots of China. The forms of these green peaks are similar to those of high-rise buildings in the city; the dimensions of the upper parts are about the same as those of the lower parts and some of the peaks have a larger upper part. The peaks are mystical displaying various poses. We can gain much enlightenment from the natural green peak forest. If we cover the surface of urban high-rise buildings or even skyscrapers with plants, we will transform the buildings into an artificial green forest where the urban buildings will have the function of absorbing CO_2 while providing oxygen; thus, the urban inhabitants will live in jeweled palaces set among enchanted hills.

Of course, there will be some new problems undertaking such transformations. The author believes that with the wisdom of mankind these problems can be solved. The main problem is a significant increase in the load on the building. If existing construction materials are used, certain elements, particularly columns, will have a much larger cross section, leading to a decrease in service area. With an increase in the height of the building, this may reach an unacceptable degree. If kilometer compressive material with its large bearing capacity is used, this problem can be solved easily. So, kilometer compressive material may provide the basis for transforming urban buildings into green peaks. The author suggests that in the future, new projects should take into account making a whole building green.

11.4.6 Other areas of application

Apart from the prospective fields of application mentioned above, the very high bearing capacity of the kilometer compressive material means it could also be used for construction of super-large-span underground tunnels, super-large-span sports buildings, anti-blasting doors, and other special buildings. Because the cross-sectional area and mass of the structural elements of kilometer compressive material can be significantly reduced to lighten the building and increase the usable area, kilometer compressive material may have wide applications in ordinary industrial and civic buildings and will gain a large market share.

In reviewing the applications mentioned above, integration of ultra-high-strength or super-high-strength concrete in a steel tube, in particular, with a super strength steel tube, is a basic technical way to prepare the kilometer compressive material under present conditions. A study shows that for such kilometer compressive material, there is ample supply of raw materials available. The preparation is simple, deformation properties are excellent, and the bearing capacity is high and can meet the requirements for construction of kilometer-high buildings. The emergence of kilometer compressive material provides a new beginning for building materials and even building technology. In the future, super-high-strength concrete and its kilometer compressive material will play an invaluable role promoting development of human society.

References

1. Wu Xiaoquan and Wen Derong, A Prosperous Year for the Development of the Concrete Industry in China, *Concrete*, 2004 (7) (in Chinese).
2. Zhou Shiqiong, *Building Materials*, Beijing: China Railway Press, 1999 (in Chinese).
3. Pu Xincheng, On the Extra Durability of the Concrete Project. *Concrete*, 2000 (1) (in Chinese).
4. Edward G. Nawy and P. E. C. Eng, *Fundamentals of High Strength High Performance Concrete*, London: Longman Group Limited, 1996.
5. Toshio Saito, Super Durable Concrete, translated into Chinese by Chen Jianxiong, *Foreign Architectural Science*, 1994 (2) (in Chinese).
6. Dai Xianming, Present Situation and Development of Concrete Industry in China, *Concrete*, 2001 (9) (in Chinese).
7. Gao Changming, Environment Friendly Characteristics of Cement and Concrete, *Concrete and Cement Products*, 2003 (2) (in Chinese).
8. P. K. Mehta, *Structure, Properties and Materials of Concrete*, translated from English by Zhu Yongnian et al., Shanghai: Tongji University Press, 1991 (in Chinese).
9. Yves Malier, High Performance Concrete (from Material to Structure), London: E & FN SPON, 1992, 270.
10. Wu Zhongwei and Lian Huizhen, *High Performance Concrete*, Beijing: China Railway Press, 1999 (in Chinese).
11. Feng Naiqian, *High Performance Concrete*, Beijing: China Building Industry Press, 1996 (in Chinese).
12. L. Pliskin, High Performance Concretes—Engineering Properties and Code Aspect, in *High Performance Concrete—From Material to Structure*, London: *E & FN SPON*, 1992, 186.
13. Odd E, Gjorv, High Strength Concrete, in *Advances in Concrete Technology*, CANMENT, 1992.
14. Pu Xincheng, Pu Huaijing, Wang Yongwei, and Wang Cong, Preparation and Study of Properties of Kilometer Compressive Material, *Concrete*, 2003 (5) (in Chinese).
15. V. D. Gluhovskii, Alkali-Slag Binder Concrete, *Concrete and Reinforced Concrete*, 1975 (3) (in Russian).

16. V. V. Ilyushin et al., *Calcium Hydrosilicate—Mono Crystal Synthesis and Crystal Chemistry*, Chinese translation from Russian by Pu Xincheng. Moscow: Nauka, 1979.
17. V. V. Timashev, Influence of Physical Structure of Cement Paste on Its Strength, *Cement*, 1978 (2) (in Russian).
18. Kuatbaev, K. K., Hydration and Hardening of the Cement (2), in Proceedings of the 6th International Congress of Cement Chemistry, Vol. II, Beijing: China Building Industry Press, 1981, 194 (in Chinese).
19. Chongqing Institute of Architecture and Civil Engineering and Nanjing Institute of Technology, *Science of Concretes*, Beijing: China Building Industry Press, 1983 (in Chinese).
20. Y. Yambor, Discussion on the 5th International Congress of Cement Chemistry, in Report on the 5th International Congress of Cement Chemistry, Stroiizdat, 1973 (in Russian).
21. P. V. Krivenko, *Durability of Alkali-Slag Concrete*, Budivelnik, 1993 (in Russian).
22. S. Nagataki, Production and Use of Super High Strength Concrete, Chinese translation from Japanese by Chen Jianxiong, *Foreign Architectural Science*, 1994 (1).
23. Ma Jianxin et al., Preparation of Plastic Highly-Flow High Strength Concrete with Water Reducing Type of Fly Ash and Slump Loss Control, *Concrete*, 1997 (1) (in Chinese).
24. Zhang Xiong et al., Characteristics and Mechanism of Action of the Composite Material from High Strength Concrete and Slag, *Concrete and Cement Products*, 1997 (3) (in Chinese).
25. Pu Xincheng, Yan Wunan, Wang Cong et al., Contribution of the Silica Fume to the Strength and Fluidity of the 150 MPa Super High Strength High-Flowing Concrete, *Concrete and Cement Products*, 2000 (1) (in Chinese).
26. Shi Yunxing et al., Surface Physico-Chemical Effect of Micro Powder in Fresh Concrete Mix, *Concrete*, 1999 (5) (in Chinese).
27. Wang Shaodong, Application of C100 High Strength Concrete in Reinforced Concrete Engineering, *Journal of Chongqing Jianzhu University*, 1999 (1) (in Chinese).
28. Cao Jianguo, Li Jinyu et al., Study on Frost Resistance of High Strength Concretes, in Proceedings of the 3rd Symposium on High Strength and High Performance Concretes and Their Application, Jinan, 1998 (in Chinese).
29. Pu Xincheng et al., Study on Fly Ash Cement with Large Addition, in *Housing Materials and Application*, 1996 (3) (in Chinese).
30. Pu Xincheng, Numerical Analysis on Pozzolanic Effect of High Strength and High Performance Concretes, *Concrete*, 1998 (6) (in Chinese).
31. Pu Xincheng, Yan Wunan, Wang Cong et al., Super High Strength High Performance Fly Ash Concrete and Analysis on Its Pozzolanic Effect, *Fly Ash*, 2000, 1 (in Chinese).
32. Pan Ganghua and Sun Wei, Influence of Super Fine Fly Ash and Silica Fume on the High Performance Concrete, *Concrete*, 1996 (3) (in Chinese).
33. Pu Xincheng, Gan Changcheng et al., Investigation and Prospect of Application of Alkali–Slag Concrete, *Concrete and Cement Products*, 1987 (5) (in Chinese).

34. Academy of Building Materials, *Test on Physical Properties of the Cement*, Beijing: China Railway Press, 1985 (in Chinese).
35. Pu Xincheng, Yan Wunan, Wang Cong et al., Technology of Preparation of 100~150 MPa Super High Strength High Performance Concrete, *Concrete and Cement Products*, 1998 (6) (in Chinese).
36. Pu Xincheng, Analysis on Strength Composition of Super High Strength High Performance Concrete, *Concrete and Cement Products*, 1999 (1) (in Chinese).
37. Ding Dajun and Jiang Yongsheng, *Introduction to Civil Engineering*, Beijing China Building Industry Press, 2003 (in Chinese).
38. Peng Xiaoping, "Low Heat High Performance High Belite Cement Concrete for Dams," doctorial dissertation, Chongqing: Chongqing University, 2002 (in Chinese).
39. Lei Fuheng, Retrospect of the Study and Application of Grade 500, Grade 1000 Harsh Concretes in China, in Papers on Application of High Strength Concretes in Engineering, Beijing: Tsinhua University Press, 1998 (in Chinese).
40. Chen Jiankui, *Principles and Application of Concrete Admixtures*, Beijing: China Planning Press, 1997 (in Chinese).
41. P. V. Sviridov, 150 MPa Concrete with Ordinary Portland Cement, *Cement and Concrete*, 1990 (2) (in Russian).
42. A. M. Neville, *Concrete Properties*, Chinese translation by Li Guoban et al., Beijing: China Building Industry Press, 1983.
43. Yu. S. Volkov, Application of Super Strength Concrete in Construction, *Concrete and Reinforced Concrete*, 1994 (3) (in Russian).
44. ACI 363 Committee, *State of the Art of High Strength Concrete*, 1994.
45. Shen Peirong, Lu Zhongfu, and Zhang Zhongqi, Study and Application of C50 Super Fine Sand Pumping Concrete, *Concrete*, 1997 (3) (in Chinese).
46. Pu Xincheng, Yan Wunan, Wang Cong et al., Technology of Preparation of Super Fine Sand Super High Strength High Performance Concrete, *Journal of Chongqing Jianzhu University*, 1999 (1) (in Chinese).
47. M. I. Netkachev, Application of Fine Sand in Concrete, in Papers of the Meeting on Concrete Technology Theory, Press of Armianian SSR Academy of Science, 1956, 319 (in Russian).
48. P. V. Sviridov, Super High Strength Cement Concrete, translated from Russian by Zhang Guomin, *Foreign Architectural Science*, 1994 (2) (in Chinese).
49. Pu Xincheng, Yan Wunan, Wang Cong et al., Study on Lithium Slag Super High Strength High Performance Concrete, in Proceedings of the 4th Symposium on High Strength and High Performance Concretes and Their Application, Changsha, 2001 (in Chinese).
50. A. M. Neville, *Properties of Concrete,* 4th ed., London: Longman, 1995, 541.
51. H. Justnes et al., High Strength Concrete Binders Part A: Reactivity and Composition of Cement Paste with and without Condensed Silica Fume, in Proceedings of the Fourth International Conference, Istanbul, 1992, Vol. II.
52. E. J. Sellevold, H. Justnes et al., High Strength Concrete Binders Part B: Non-Evaporable Water, Self-Desiccation and Porosity with and without Condensed Silica Fume, Ash, Silica Fume, Slag and Natural Pozzolans in Concrete, in Proceedings of the Fourth International Conference, Istanbul, 1992, Vol. II.

53. Wan Chaojun, "Study on New Effective Additive and Its Properties," doctorial dissertation, Chongqing: Chongqing University, 2000 (in Chinese).
54. Wang Yongwei, "Study on Composition, Structure and Shrinkage and Its Compensation of Super High Strength High Performance Concretes," master's dissertation, Chongqing: Chongqing University, 2001 (in Chinese).
55. H. F. W. Taylor, *Cement Chemistry*, Academic Press, 1995.
56. A. E. Seikin et al., *Structure and Properties of Cement Concrete*, Chinese translation from Russian by Hu Chunzhi, Yuan Xiaomin et al., Beijing: China Building Industry Press, 1984 (in Chinese).
57. Pu Xincheng and Wang Yongwei, Influence of Effective Active Mineral Additive on the Pore Structure and Interface Structure of Super High Strength Concrete, 2003 Academic Conference of China Ceramic Society: Papers on Cement-Based Materials (Part 2), Beijing, 2003 (in Chinese).
58. Lu Ping, *Introduction to Science of Cement Materials*, Shanghai: Tongji University Press, 1991 (in Chinese).
59. Xu Zhongzhi and Shen Yang, Strength Performance and Pore Structure of Mortar with Low Porosity, *Journal of Nanjing University of Chemical Technology*, 1994 (16) (in Chinese).
60. U. Nilsen, P. Sanderg, and K. Follard, Influence of Mineral Admixtures on the Transition Zone in Concrete: Interfaces in Cementitious Composites, in Proceedings of the International Conference held by RILEM, London: E & FN SPON, 1993:65.
61. Chen Zaoyuan, Zhu Jinquan, and Wu Peigang, *High Strength Concrete and Its Application*, Beijing: Tsinhua University Press, 1992 (in Chinese).
62. Xu Jinfeng, Zhuang Yaping et al., Study on Whole Curve of Stress Strain of High Strength Concrete, in Papers of the Symposium on Theory of Strength and Application of Restricted Concrete and Ordinary Concrete, Yantai, 1987 (in Chinese).
63. Jiang Zhengwu, Sun Zhengping, and Wang Peiming, Study on Relative Moisture Variation of High Performance Concrete Itself, *Journal of Ceramics*, 2003 (8) (in Chinese).
64. O. M. Jesen, Thermodynamic Limitation of Self Desiccation, *Cement and Concrete Research*, 1995, 1.
65. Bian Rongbing, Synthesis and Test of Contraction Agent for Concrete, *Chemical Building Materials*, 2002 (5) (in Chinese).
66. Hiroshi Muguruma, Expansive Cement Concrete, *Concrete Journal*, 1974 (11) (in Japanese).
67. Gan Changcheng et al., Study and Application of Low Strength High Impermeable Concrete, *Concrete*, 2003 (12) (in Chinese).
68. Han Jianguo and Yang Fumin, Mechanism and Effect of Application of Contraction Agent for Concrete, *Concrete*, 2001 (4) (in Chinese).
69. Yuichi Otabe, Development of Initial Defect-Free High Performance Concrete, *Cement & Concrete*, 2001 (658, No. 12), pp. 36–44 (in Japanese).
70. T. Komuro, S. Kuroiwa, H. Watanabe, and H. Jinnai, Research for Application of RC Columns Using Super High Strength Concrete with Compressive Strength of 150 N/mm^2, *Concrete Journal*, 2001 (39, No. 10), pp. 9–16 (in Japanese).
71. S. P. Shah and S. H. Ahmad, *High Performance Concrete and Applications*, London: Edward Arnold, 1994.

72. Li Jin yu and Cao Jian guo, *Research on Durability and Application of Hydraulic Concrete*, Beijing: China Electric Power Press, 2004 (in Chinese).
73. Pu Xincheng and Zhao Zhenhao, *Building Products from Lime-Sand Silicates*, Beijing: China Building Industry Press, 1980 (in Chinese).
74. A. M. Brandt, *Cement-Based Composites: Materials, Mechanical Properties and Performance*, London: E & FN SPON, 1995, 320.
75. Toru Kawai, The Super High Strength Concrete—Challenge of Utilization of Concrete with Compressive Strength of 1000kg/cm^2, *Cement & Concrete*, 1986 (6) (in Japanese).
76. J. Kropp and H. K. Hisdort, *Performance Criteria for Concrete Durability* (RILEM Report 12), London: E & FN SPON, 1995, 193.
77. M. Roller, Physical, Mechanical and Durability Characteristics of Various Compressive Strength Concretes (from 20 to 120 MPa), in Proceedings of CANMENT/ACI International Conference on Advances Concrete Technology, CANMENT, Athens, Greece, 1992.
78. Feng Naiqian, *High Strength Concrete Technology*, Beijing: China Building Materials Industry Press, 1992 (in Chinese).
79. Qian Xiaoqian and Zan Shulin, Strength and Inspection of High Strength Concrete Specimen, *Concrete*, 2000 (6) (in Chinese).
80. High Strength Committee and China Civil Engineering Association, *Guide to Structural Design and Construction for High Strength Concrete*, 2nd ed., Beijing: China Building Industry Press, 2001 (in Chinese).
81. Wang Yurong, Preparation and Experimental Study of C80 High Performance Pumping Concrete, *Concrete*, 2000 (6) (in Chinese).
82. Pu Xincheng, Yan Wunan, Wang Cong et al., Study and Prospect of Application of 150 MPa Super High Strength High Performance Concrete, *Concrete*, 2003 (7) (in Chinese).
83. Wu Peigang et al., Research and Application of High Strength Concretes, in Papers on Application of High Strength Concretes in Engineering, Beijing: Tsinhua University Press, 1998 (in Chinese).
84. Yan Wunan, Pu Xincheng, Wang Cong et al., Study on Chemical Shrinkage and Drying Shrinkage of Super High Strength Concrete, *Construction Technology*, 1999 (6) (in Chinese).
85. I. Holland (ed.), Application of High Performance Concrete. Report of the CEB-FIP Working Group on High Strength/High Performance Concrete. *CEB Bulletin d'Information*, No. 222, 1994.
86. Pu Xincheng, Research and Application of Super High Strength Concretes, *Concrete*, 1993 (5) (in Chinese).
87. A. M. Brandt, *Cement-Based Composites: Materials, Mechanical Properties and Performance*, London: E & FN SPON, 1995, 430.
88. Beijing Urban Construction Group Co. Ltd (ed.) *Practical Handbook of Building Construction Cases*, Beijing: China Building Industry Press, 1999 (in Chinese).
89. Lin Liyan, Lyu Yudan, and Lin Gan, Integration of Concrete with Steel for Promoting Development of High-Rise Building Structures, in Proceedings of 4th Symposium on High Strength and High Performance Concretes and Their Application, Changsha, 2001 (in Chinese).

90. Xu Xin, Han Sufang, and Yu Dazhong, Application of C80~C100 High Performance Concrete in Engineering, in Papers of the Symposium on High Strength and High Performance Concretes and Their Application, Fuzhou, 2002 (in Chinese).
91. Lu Laijun and Zhu Xiaorong, Research and Application of C100 High Performance Concretes, *Concrete*, 2003 (7) (in Chinese).
92. Guo Peiling, Shi Dongqing et al., Application of C100 Super High Strength Concrete in the Project of Shenyang Yuanji Building, *Concrete*, 2003 (7) (in Chinese).
93. Ekkehard Fehling et al., Ultra High Performance Composite Bridge across the River Fulda in Kassel, in Proceedings of the International Symposium on Ultra High Performance Concrete, Kassel, Germany: Kassel University Press, 2004.
94. Urs Maeder et al., Ceracem: A New High Performance Concrete: Characterisations and Applications, in Proceedings of the International Symposium on Ultra High Performance Concrete, Kassel, Germany: Kassel University Press, 2004.
95. Ziad Hafar et al., Design and Construction of the World's First Ultra-High Performance Concrete Road Bridges, in Proceedings of the International Symposium on Ultra High Performance Concrete, Kassel, Germany: Kassel University Press, 2004.
96. Gao Xueshan, *Something about Concrete*, Beijing: China Building Industry Press, 1985 (in Chinese).
97. J. Webb, High Strength Concrete: Economics, Design and Ductility, *Concrete International*, 1993 (1).
98. Zhang Lide and Mou Jimei, *Nanometer Materials and Nanometer Technology* (M), Beijing: Science Press, 2001 (in Chinese).
99. Tan Kefeng, Pu Xincheng, and Cai Shaohuai, Properties and Ultimate Bearing Capacity of Steel Tube Super High Strength Concrete, *Journal of Building Construction*, 1999 (1) (in Chinese).
100. Cai Shaohuai, *Structural Calculation and Application of Steel Tube Concrete*, Beijing: China Building Industry Press, 1989 (in Chinese).
101. Cai Shaohuai, Recent Advancement of Technology for Steel Tube High Strength Concrete Structures in China, *Journal of Civil Engineering*, 1999 (4) (in Chinese).
102. Pu Xincheng, Pu Huaijing, Wang Yongwei, and Wang Cong, Kilometer Compressive Material and Approaches to Its Preparation, in Papers on Advancement of Materials Science and Engineering in 2002, Beijing: Metallurgy Industry Press, 2003 (in Chinese).
103. Wang Yongwei, "Research on Preparation and Mechanical Performances of Kilometer Load-Bearing Material," doctorate dissertation, Chongqing: Chongqing University, 2004 (in Chinese).
104. Xiang Haifan, Prospects for Bridge Engineering of the World in 21st Century, *Journal of Civil Engineering*, 1999 (3) (in Chinese).
105. Yang Boke, *Practical Guide for Concrete*, Changchun: Jilin Press of Science and Technology, 1998 (in Chinese).

106. Pu Xincheng and Wang Yongwei, Influence of Effective Active Mineral Additive on the Hydration Degree and Characteristics of Hydrate Composition, 2003 Academic Conference of China Ceramic Society: Papers on Cement-Based Materials (Part 2), Beijing, 2003 (in Chinese).
107. S. A. Podmasova, S. H. T. Balev, and Y. D. S. Volkov, New Low Water Demand Binders for High Strength Concretes, in Proceedings of International Conference on High Strength Concrete, Norway, 1993.
108. Li Qiang, Zhao Mingzhe, and Li Congzhi, Synthesis and Property Analysis of Superplasticizers from Ammonia-Based Sulfonates, *Concrete*, 2001 (11) (in Chinese).
109. Wang Cong, Pu Xincheng, Yan Wunan et al., Study on Resistance of High Performance Concrete to Chemical Corrosion, *Journal of Chongqing Jianzhu University*, 1999 (1) (in Chinese).
110. Pu Xincheng, Wang Yongwei, Pu Huaijing, and Wang Cong, Kilometer Compressive Material and Approaches to Its Preparation, *Journal of Civil Engineering*, 2004 (7) (in Chinese).
111. Xincheng Pu, Chaojun Wan, Yong Wei Wang et al., Compressible Material and Its Preparation in Proceedings of the International Symposium on Ultra High Performance Concrete, Kassel, Germany: Kassel University Press, 2004.

Index

A

B

D

N

O

P

Q

R

S

T

Y

Z

For Product Safety Concerns and Information please contact our EU representative GPSR@taylorandfrancis.com
Taylor & Francis Verlag GmbH, Kaufingerstraße 24, 80331 München, Germany

www.ingramcontent.com/pod-product-compliance
Lightning Source LLC
LaVergne TN
LVHW011120300926
849160LV00008B/452
9781138075641